The first step in identifying any plant specimen is the identification of the family to which it belongs. This short, user-friendly guide provides all the information necessary to identify correctly the flowering plant families found normally either in the wild or grown in cultivation (outside or under glass) in northern temperate regions. In total, 286 families are covered by the bracketed identification keys, which are accompanied by a comprehensive explanatory guide to their use. A fully illustrated discussion of floral structure and terminology precedes the keys and brief descriptions of the families, arranged for convenience according to the Engler and Prantl taxonomic system, follow. A chapter giving advice on identification beyond the level of family, plus an annotated bibliography and glossary, complete the volume.

In this new edition the information on morphology and terminology has been restructured, the keys have been revised and improved, and the descriptions rewritten and expanded so that they are more accessible and informative. Aimed primarily at students of botany and horticulture, this volume will also serve as a basic introduction for all those with an interest in plant identification.

THE IDENTIFICATION OF FLOWERING PLANT FAMILIES

FOURTH EDITION

THE IDENTIFICATION OF FLOWERING PLANT FAMILIES

including a key to those native and cultivated in north temperate regions

J. CULLEN D.SC.
Director, Stanley Smith (UK) Horticultural Trust

FOURTH EDITION
completely revised and edited

CAMBRIDGE
UNIVERSITY PRESS

PUBLISHED BY THE PRESS SYNDICATE OF THE UNIVERSITY OF CAMBRIDGE
The Pitt Building, Trumpington Street, Cambridge CB2 1RP, United Kingdom

CAMBRIDGE UNIVERSITY PRESS
The Edinburgh Building, Cambridge CB2 2RU, United Kingdom
40 West 20th Street, New York, NY 10011-4211, USA
10 Stamford Road, Oakleigh, Melbourne 3166, Australia

First published by Oliver & Boyd 1965

Second edition published by the Cambridge University Press 1979
Reprinted 1980
Third edition 1989
Reprinted 1990, 1994
Fourth edition 1997

Printed in the United Kingdom at the University Press, Cambridge

Typeset in Linotype Lectura 8½/12 pt

A catalogue record for this book is available from the British Library

Library of Congress Cataloguing in Publication data

Cullen, J. (James)
    The identification of flowering plant families : including a key
to those native and cultivated in north temperate regions / J.
Cullen. − 4th ed.
        p. cm.
    Completely rev. and edited ed. of: The identification of flowering
plant families / P. H. Davis, J. Cullen. 1989.
    Includes bibliographical references and index.
    ISBN 0 521 58485 X (hb). − ISBN 0 521 58550 3 (pbk.)
    1. Angiosperms−Identification.   2. Flowers−Identification.
I. Davis, P. H. (Peter Hadland). The identification of flowering
plant families.   II. Title.
QK495.A1C85   1997
582.13′0912′3−dc20   96−41206   CIP

ISBN 0 521 58485 X hardback
ISBN 0 521 58550 3 paperback

Contents

Illustrations

Preface

The process of improvement and adaptation to the requirements of the times, which have been a feature of the development of this book over three previous editions, have continued in this, the fourth edition. The main changes over this thirty year period were made between the second and third editions, and the preface to the third edition is reprinted here (pp. x–xi) for convenience.

The main changes from the third edition to the present one are as follows:

(a) The two sections of the third edition headed 'Usage of terms' and 'Examining the plant' have been amalgamated and considerably extended to produce a new section (pp. 4–44) on flowering plant morphology as used in the keys and descriptions.

(b) The keys have been improved and the terminology used in them has been revised.

(c) The section entitled 'Arrangement and description of families' has been completely re-written. The abbreviations which were used extensively in the earlier editions have almost all been removed, so that the descriptions are easier to read and understand, and the notes on the various families are more extensive, and, I hope, more useful.

(d) The glossary has been extended to accommodate the extra information introduced under (a) above.

James Cullen
Cambridge, May 1996

Preface to the third edition

That a book of this type should appear in a third edition implies that it has a certain usefulness and practicability. The main purpose of the changes made to the text in preparing this edition has been to enhance and improve these qualities.

Since 1979, when the second edition was published, numerous changes have taken place in taxonomic thinking and in public attitudes to plants and wildlife generally. Some of these changes continue the trends noted between the first (1965) and second editions, particularly the continuing decline of the teaching of plant taxonomy in the Universities and the increasing separation of professional plant taxonomists from the public that needs the results of their work.

On the other hand, there have been some encouraging developments as well, especially the increasing public awareness of the need for conservation, which has led to a renewed interest in plants (and animals). This interest is in part satisfied by excellent wildlife programmes on television and by the numerous, well-illustrated, popular books on plants and animals available today. Less, however, is done for those who have a more serious, practical interest, and wish to know more about the scientific background. This edition is intended for such people, whether they are botanists, gardeners, landscape architects or otherwise involved with plants, either as students, professionally, or as a hobby.

The main changes from the second edition are as follows:

(a) The adoption of the Engler & Prantl taxonomic system as the basis for the recognition of families and their taxonomic sequence. This may appear regressive to professional taxonomists, as many more up-to-date systems have been published since 1979. However, the Engler & Prantl system still seems to be the best one for general purposes: it is well documented and widely used; its use in *The European Garden Flora* (1984 and continuing) gives further impetus for its use here.

(b) The keys have been modified to take account of errors and difficulties in use that have come to light since 1979 and the section 'Using the keys' has been enlarged to provide further guidance.

(c) The section 'Further identification' has been completely re-written.

(d) Many minor changes, especially in terminology, have been incorporated.

Finally, the last paragraph of the preface to the second edition remains valid, and I quote it without apology: 'In our view a natural classification of plants and their correct identification (for which the families are an important step on the road) remain essential for the progress of biology on a broad front.'

James Cullen
Edinburgh, July 1988

Acknowledgements

Over the years of this book's existence, many botanists and others have helped with advice and information on matters of detail. These are now too numerous to mention individually, and include several correspondents who have pointed out errors or omissions. I hope they will accept this general acknowledgement. In the preparation of this fourth edition I am particularly grateful to Sabina Knees and Suzanne Maxwell for both botanical and general assistance. The illustrations, reprinted from the second edition, are by Rosemary Smith.

Introduction

The identification of the family to which a plant belongs is the first step in its complete identification. Because the number of families in the flowering plants is relatively small (between 350 and 520, depending on whose system is followed), it is theoretically possible to know them all (something not possible for genera or species); this is rarely achieved, however, by any individual.

This key attempts to provide a means of identification for all flowering plant families native, or cultivated (out-of-doors or under glass) in north temperate regions. In practice, the southern limit of the area involved is approximately 30° N, thus excluding all of Mexico and Florida in the New World and most of India and tropical China in the Old World. A few, mainly tropical families with a small number of genera native in China north of the limit have also been excluded as, unless cultivated in Europe or North America, they are so infrequently seen. The key (with sub-keys included with the descriptions of some larger families) covers 286 families.

As far as cultivated plants are concerned, a few tropical families which rarely flower in cultivation have been included (e.g. *Dipterocarpaceae*), as they may occasionally be seen in botanic garden collections. The key has been constructed to allow for the identification of the frequently cultivated representatives of tropical and southern hemisphere families; it may not work for other genera. No attempt has been made to allow for the double-flowered, wilder excesses of the plant-breeder.

The family is merely one level of the taxonomic hierarchy, but it is generally the highest level that has significance in identification (but see p. 46 for details of the Monocotyledons and Dicotyledons). Families contain genera, which may themselves contain species. A family may consist effectively of only a single species, e.g. Scheuchzeriaceae, which comprises the single genus *Scheuchzeria*, itself made up of the single species *Scheuchzeria palustris*. On the other hand, large families contain some hundreds of genera and several thousand species (e.g. *Orchidaceae, Compositae, Leguminosae, Rubiaceae*). This variation in size means that some families are considerably more variable than others, and that characters that are generally diagnostic in some cases are not so in others; hence many families key out more than once in this key. The variability also means that what constitutes a family is to a considerable extent a matter of opinion, and in recent years there has been a strong tendency to split up the large families into more numerous segregate families. Here a 'middle way' is adopted by following a well-established, conservative taxonomic system (that of Engler & Prantl as expressed in *Syllabus der Pflanzenfamilien*, edn 12, vol. 2, ed. H. Melchior), while indicating, by means of sub-keys under the brief descriptions of the families (pp. 91–187) some of the more widely recognised segregates.

Families are grouped together in terms of what are considered their affinities or relationships into orders and higher groups. In this book the orders (whose names end in '-ales') are briefly mentioned in the section on family descriptions, but the higher groups (apart from the Monocotyledons and Dicotyledons) are not mentioned.

An arrangement of families in their orders (and higher groups) into a particular linear sequence forms a taxonomic system. The sequence of families is used by authors of taxonomic systems as a means of indicating their ideas about how the flowering plants have evolved. There are many differing systems available at present in the taxonomic literature, though none of them has won particularly wide acceptance (the differences between the various systems are often trivial). That followed here is the most generally useful for identification, being widely followed in Floras (e.g. *Flora Europaea*, *The European Garden Flora*), well-documented and widely available. Its adoption here is for convenience and practicality and implies no particular view of the evolution of the flowering plants.

The short descriptions of the families are intended both as a check on identification and as a terse presentation of the important family characteristics. These descriptions refer to the families as wholes, not just to those representatives covered by the key. The distribution of each family is given, though without great detail. For families consisting of only a single genus, the name of that genus is also given.

The long chapter on plant structure (Examining the plant — p. 4) provides a brief survey of plant structure and its associated terminology, as used in the key and family descriptions. This should be studied carefully by inexperienced users.

The short section entitled 'Further Identification and annotated bibliography' (p. 188) is intended to help the user to proceed further with the identification process. This section covers only the main literature useful for plant identification.

It is not possible, in a book of this size, to give information about the theory and practice of plant classification. It should be remembered that the name of a plant is only a key to the information available about it, not an end in itself. Any user whose interest in this aspect of the subject is stimulated by the practice of identifying plants is recommended to consult a good library where botanical books are well represented.

Examining the plant: a brief survey of plant structure and its associated terminology

The identification of plants is carried out on the basis of the information available about the plant in each particular case. In most situations, this information will be derived from a specimen of the plant itself, either whole (if the plant is small, or if it is being examined while growing) or a part (generally a stem or twig, with or without flowers or fruits), and consists of the structure displayed by the specimen (its morphology) together with other information that may be available (e.g. where the plant came from originally). On the basis of this information, one can make use of the keys in this book to obtain an accurate identification of the family to which the specimen belongs. In order to do this, the specimen has to be observed carefully, so that the structure it displays, and the terminology needed to describe it, are properly understood. The rest of this chapter gives a very brief survey of flowering plant morphology in so far as it is needed for family identification. Each new term is italicised at its first appearance, and appears in the glossary (p. 195). Further information can be found in textbooks of botany, and in Bell, A. D., *Plant Form*, Oxford (1991), which is extremely well illustrated, with fine photographs.

The level of detail included here covers what can be seen with the naked eye or with the aid of a hand-lens magnifying 10 or 15 times, or other directly perceptible characteristics (e.g. scent). In making classifications, plant taxonomists may use not only these characteristics but also others which require more complex equipment – microscopes (both light and electron microscopes), various pieces of apparatus, etc. The classifications so produced, however, are generally expressible at the simple naked-eye morphological level, even though their information-base is much wider than this.

[4]

1 Duration and habit

Plants may be *herbaceous*, that is, they produce little or no persistent, woody tissue above ground and their stems are soft and without obvious bark, or *woody*, with persistent, hard, aerial twigs which usually possess obvious bark.

Herbaceous plants may persist for just one growing season: the seed germinates, grows into a plant which produces flowers, fruits and seeds and then dies off, all within the one continuous span. Such plants are known as *annuals*. In north temperate areas, most annuals germinate in the spring and die off in the autumn or early winter; a few, such as *Arabidopsis thaliana* (Thale cress) or *Capsella bursa-pastoris* (shepherd's purse), both fairly common garden weeds, germinate in autumn, pass the winter as small rosettes of leaves near the ground, and flower in the following spring: such plants are known as *winter-annuals*. Annuals may be recognised by the following features: they have small, slender roots (often surprisingly small for the bulk of the plant above the ground), and almost all the branches produce flowers or inflorescences, particularly towards the end of the growing season.

Herbaceous plants which last for two seasons are called *biennials*: they usually germinate in the spring and produce a rosette of leaves during the first year, which persists through the subsequent winter and then produces flowering shoots, fruits and seeds during the following spring or summer, following which the whole plant dies off. As in annual plants, all or most of the shoots eventually produce flowers and fruits. The distinction between annuals and biennials is not always clear-cut, especially with plants seen on only one occasion in the wild (in a garden, of course, plants can theoretically be observed throughout their life-cycles). Biennials can, however, usually be recognised by the co-existence of 1-year-old non-flowering rosettes growing amongst flowering plants of the same kind.

There are some plants which act like biennials in that they produce a rosette of leaves which does not flower immediately; this rosette may persist for several years, 5 or 6 in the case of some species of *Meconopsis* (the Himalayan poppy), 50 to 100 in the case of some species of *Agave*. Such plants are described as *monocarpic*.

Herbaceous plants which persist for several seasons, flowering every year (except sometimes their first), are called *herbaceous perennials*. Their flowering stems die back to ground-level every winter, and the plant persists (*perennates*) as underground parts, which can become quite woody. Occasionally, in some species of herbaceous perennials, leaf-rosettes persist at ground-level through the winter. In all, however, some shoots in each year do not produce flowers and fruits, but form the basis of growth for the subsequent season.

Woody plants have aerial, woody stems and twigs which persist through several to very many winters. The shoots may be thin and wiry or thick and massive, but whatever their size, they bear buds which allow for further growth during spring and summer, and often have noticeable bark (in plants from areas where growth is possible throughout the year, buds as such are strictly not present, the growing points producing new leaves as and when appropriate).

Subshrubs are generally small, low plants with thin, wiry, woody stems; they can be easily mistaken for herbaceous perennials, but are distinguished by the persistent, woody shoots, as seen, for example, in many species of heather (*Erica* species). In Latin, such plants are known as *suffrutices*, and the adjective derived from this, *suffrutescent*, is sometimes used in the botanical literature.

Shrubs are larger, woody plants with obvious, persistent branches. Generally, they have several main stems rather than a single trunk, and these main stems tend to be branched from near ground-level. There is no sharp distinction between shrubs and *trees*; the latter are generally larger than shrubs, and usually have a distinct *trunk* or *bole* (sometimes several) which tends to raise the branches well above ground-level (though trunks may well branch nearer the base when the plant is older). These usages of the terms shrub and tree correspond fairly well with the terms as used in common speech, but the degree of precision is somewhat greater.

A few woody plants behave *monocarpically* (see above) in that they build up not a rosette of leaves, but a plant-body that is a tree or shrub, which bears flowers only once, and then the whole plant dies, at least to the level of the underground parts (flowering may last for several years). Such behaviour is described as *hapaxanthic*, and is seen in some species of palm.

Climbers may be either herbaceous or woody. A few (mainly tropical) plants can be shrubs if no support is available, or climbers if it is available; such species, if support becomes available during the lifetime of the plant, can begin as a shrub, continue as a climber and finally succeed as a tree.

A small number of plants are *parasitic* on other plants; that is, they draw all or most of their nutrition from the host plant; such parasites tend to have very reduced plant-bodies, lack chlorophyll and generally have a rather simplified vegetative morphology. They should not be confused with *epiphytes*, plants which grow on other plants without extracting any nourishment from the plants they grow on. Epiphytes tend to have 'normal' (i.e. not very reduced) vegetative morphology. A few plants are *half-parasites* in that they draw nutrients from host plants but also support themselves to some extent by photosynthesis (e.g. species of *Melampyrum*). Again, a small number of plants are *saprophytes*, absorbing complex chemicals from the soil and its fungal components rather than making them themselves through photosynthesis.

II Underground parts

These are not extensively used in plant identification, because they are not often seen, but their importance in the distinction between the various kinds of herbaceous plants has already been noted. There are several distinct types of underground part.

Roots anchor the plant in the soil and absorb water and minerals. They generally grow downwards or downwards and outwards, are never green, and never bear leaves or buds. The first root of the seedling, the primary root, may persist, growing in length and thickness and bearing many branches, forming a taproot system (as in most Dicotyledons), or the primary root may not last long, its functions being taken over by roots produced from buds at the bases of the stems (*adventitious* roots), forming a fibrous root system (as in most Monocotyledons). Some plants bear roots which become swollen and act as food-storage organs (e.g. the carrot); such organs are known as *root-tubers*.

Some plants, mainly those that grow epiphytically (i.e. on other plants) produce aerial roots from adventitious buds on the stems. These roots may descend to the soil where they absorb nutrients and water, as in some tropical orchids and the familiar Swiss-cheese plant (*Monstera pertusa*), or they may simply hang in the damp atmosphere and absorb moisture (as in many tropical orchids). In a few cases, aerial roots have other functions (e.g. the climbing roots of ivy, *Hedera helix*).

Underground stems look superficially like roots, but they bear buds and small, reduced leaves (*scale-leaves*), and frequently grow horizontally or almost so. If they reach the light, they turn green and the buds may develop into shoots which bear leaves. Such underground stems are known as *rhizomes*; they may become swollen for food-storage, when they are known as *tubers* (as in the potato, *Solanum tuberosum*). Rhizomes occasionally extend above ground, looping and then rooting at a point some distance from the base of the parent plant; such aerial rhizomes are known as *stolons* or *runners* – the strawberry plant (*Fragaria*) provides a familiar example.

Rhizomes which are very short, swollen, bulb-like and upright are known as *corms*, as seen in the species of *Crocus*. *Bulbs* are complex organs made up of modified roots, stems and leaf-bases. Most of the bulb consists of the swollen bases of the leaves, which overlap and enfold each other (as in the onion, *Allium cepa*) and are attached at the base to a flat or broadly pyramidal plate which is the effective stem (bearing roots on its outer side). The outermost leaf-bases tend to be fibrous or papery, and serve as protection for the more delicate tissues within.

III   Above-ground parts

These are the most conspicuous parts of the plant and form what is commonly thought of as 'the plant' itself. They are attached to the root, mostly at or near soil-level by a transitional zone which is sometimes called the *stock* or *caudex*. The aerial parts may be very extensive, consisting of various organs, which will be described here serially from the base upwards.

1 Stems

These are the main supporting structures of the plant above ground, bearing the buds, leaves, flowers and fruits. They are generally *terete* – circular in section, though square sections are found in the *Labiatae*, and stems can be winged or with other outgrowths (in many climbers and a few erect species). They may be erect to horizontal, sometimes erect near the base, then arching over so that the tips are pendulous. With woody plants, the term 'stem' is rarely used, the words 'trunk', 'branch', 'twig' or 'shoot' being used depending on the size on the part in question. The term 'stem' is used here in a more precise sense than it is in general English. For instance, the stalk on which a dandelion (*Taraxacum*) flower-head is borne is technically not a stem (it bears neither leaves nor buds), though often described as such in common speech. Such stalks, which occur in many herbaceous plants, especially those with bulbs, bearing the flower(s) above a ground-level rosette of leaves, are correctly termed *scapes*.

Most stems bear a bud or growing point at the tip, and grow by means of this, bearing leaves, etc. at varying distances from each other, with leafless parts of the stem between; in some plants there is a stem-dimorphism, with some stems like those described above, others, which tend to bear the active leaves, scarcely elongating, forming condensed *short-shoots* (e.g. many species of *Berberis*).

2 Leaves

Leaves are present in most plants, and form an extremely variable set of organs; they produce many features which are important in identification. In most plants they are borne directly on the stems, twigs or branches, but they may also be borne in basal rosettes or on short-shoots (see above).

The point on the stem, twig or shoot at which a leaf, a pair of leaves or a whorl of leaves is borne is known as a *node*; the leafless parts of the shoot, between the nodes, are known as *internodes*.

Duration. Leaves may last for only a single growing season, emerging from buds in the spring and dying off and falling in the autumn; such leaves are *deciduous*. Or, they may last for several seasons, when they are *evergreen*. Deciduous leaves are usually thin and papery or parchment-like in texture, while evergreen leaves are usually thicker and somewhat leathery (*coriaceous*), or needle-like (*acicular* or *subulate*). In a few species the leaves are *half-evergreen*, some falling after one season, others persisting for longer.

Attachment to stem. The leaves are attached to the stems at the nodes. There may be a single leaf at each node, a pair, or a whorl. When there is a single leaf per node, and the leaves are arranged along an imaginary spiral, then they are described as *spirally arranged*. If the leaves are similarly 1 per node, but successive leaves are on opposite sides of the stem, then the leaves are *alternate* or *distichous* (sometimes no distinction is made between spirally arranged and alternate and both types are referred to as alternate, though this is not strictly accurate). If there are 2 leaves at each node, they are generally arranged on opposite sides of the stem, when the leaf arrangement is termed *opposite*; when the successive pairs of leaves are at 90° to each other, the arrangement is described as opposite and *decussate*. When there are several leaves at each node, the arrangement is described as *whorled*. Care must be taken to distinguish between spirally arranged leaves borne close together and forming a *false whorl* (as in many species of *Rhododendron*) and true whorls. Leaves may be borne in rosettes at the base of the flowering scapes, as in the dandelion (*Taraxacum*) and many other species. In such rosettes the internodes are very reduced and the leaves are borne very close together. It is, however, generally easy to see that the leaves in such rosettes are spirally arranged.

The leaves may be attached to the node by a long or short stalk, known as a *petiole*; this may be sharply distinguished from the broader part (the *blade* or *lamina*), or the blade may taper imperceptibly into it. Leaves which have no petiole are described as *sessile*.

The upper angle between the petiole (or the leaf-base if the leaf is sessile) and the stem to which it is attached is known as the *axil*. In each axil there is a bud or a branch which has developed from a bud; some plants have multiple buds, one above the other, in each axil (e.g. some species of *Spiraea*).

Division of the blade. The leaves of most plants are unitary structures, not divided up into separate segments or *leaflets*. Undivided leaves, which may be variously lobed or toothed, are described as *simple*, and as *entire* when they have no sign at all of lobing or toothing (i.e. the margin is smooth and unindented, as in privet, *Ligustrum vulgare*). Toothed leaves have margins which are incised slightly, either regularly or irregularly; when the toothing is sharp and regular, like that of a saw, the leaves are described as *serrate*; when each tooth is itself regularly toothed, the term *biserrate* can be used. When the teeth are larger and not so regular, the margin is described as *dentate*; dentate leaves can be either regularly or irregularly dentate. When the teeth are rounded rather than sharp, the margin is described as *crenate*. All such leaves have only the margins divided. Lobed leaves are incised to at least one-third of the distance from margin to midrib; the terminology used to describe them is related to that used for *compound* leaves (leaves divided into distinct leaflets), so these are discussed together below.

Compound leaves are made up of 2 or more quite separate segments which are called *leaflets*; leaflets may be stalked, when they are described as *petiolulate* (as in the common house-plant, *Schefflera*), or, more commonly, they are without stalks (i.e. sessile). To the superficial observer, leaflets can easily be confused with individual leaves; the essential difference is that at the base of the leaf, there is a bud (or a branch which has developed from a pre-existing bud) in the axil, whereas for leaflets neither of these organs is present. Leaves may be divided into leaflets in two ways: the leaf may be arranged like a feather, with the leaflets arranged parallel to each other along the sides of the main axis (*rachis*), which is a continuation of the petiole; or all the leaflets may arise from the same point at the top of the petiole. The first type of division is known as *pinnate*, the second as *palmate*. In pinnate leaves, the leaf may end in a single terminal leaflet (when it is described as *imparipinnate*), or there may be no obvious terminal leaflet, when the leaf is described as *paripinnate*. In some climbing plants the terminal leaflets of the pinnate leaves are replaced by tendrils (see below).

In the case of a plant which has leaves consisting of 3 leaflets only, such as most clovers (species of *Trifolium*) the leaf could theoretically be either pinnate or palmate or both; to avoid confusion, such leaves are described as *trifoliolate* (sometimes mis-spelled 'trifoliate').

In all compound leaves, the leaflets are separate from each other right to their bases. Leaves can, however, be simple but lobed; the lobing may extend only as far as one-third of the distance from margin to midrib, or it may extend almost (but not quite) to the midrib. Leaves which are lobed pinnately from one-third to two-thirds of the distance from margin to midrib are known as *pinnatifid*; if the lobing reaches further than two-thirds of this distance, but does not quite separate the blade into individual leaflets, the leaf is described as *pinnatisect*. Similarly, the terms *palmatifid* and *palmatisect* are used to describe leaves which are lobed in a manner reminiscent of palmate division.

The leaflets of divided leaves can themselves be toothed, lobed or further divided into leaflets of the second degree (and, rarely, these into leaflets of the third or even fourth degree). Such division is found only in pinnately divided leaves, which are then known as *bipinnate* or doubly pinnate (as in the florist's Mimosa, which is actually a species of *Acacia*), *tripinnate*, or *quadripinnate*.

Shape. The shapes of leaves (or, when appropriate, leaflets) are infinitely variable, and a huge terminology has been developed to cope with this variability. This terminology is based on the overall shape considered as the length relative to the breadth, and the position of the widest part (whether at the middle of the leaf or in the upper or lower thirds). It is not important at the level of family identification and so is not discussed in detail here.

Stipules. Around the point at which the petiole attaches to the stem, there may or may not be two outgrowths known as *stipules*. Some plants have them, others do not, so their presence or absence is often important in identification. If they are present, they can be extremely variable in form, sometimes very small and inconspicuous, sometimes large and leaf-like. In a few species (e.g. *Lathyrus nissolia*) they are larger than the rest of the leaf and form the main photosynthetic organs.

Stipules may be separate from the petiole, or joined to it, as in the rose (*Rosa*), when they are described as *adnate* to the petiole (the word adnate is always used to indicate that organs of differing types are joined together, in this case petiole and stipule; the word *connate* is used when organs of the same type are joined together, e.g. connate petals). When the leaves are opposite, the stipules may form a pair on each side of the stem, between the attached bases of the leaves, as in many members of the family *Rubiaceae*. In some members of this family (e.g. bedstraws, species of *Galium* and its allies) the 4 stipules belonging to each pair of leaves are as large as, and similar to the leaves in structure and appearance, so that it appears that the leaves are borne in whorls of 6 (sometimes whorls of 4 by suppression of 2 stipules); examination of the position of the buds reveals which of the 6 are genuine leaves.

Stipules, when present, may be very quickly deciduous, falling almost as soon as the leaves have expanded. This happens in many tree species (e.g. *Betula*, *Quercus*), and the bases of the mature leaves must be examined very carefully to see the small scars left by the fallen stipules. In *Magnolia* each young leaf in bud is completely wrapped by its stipules, which provide protection against severe weather.

Leaf-scars. The importance of observing the scars left by fallen organs has already been mentioned above. In woody plants, scars are also developed when leaves fall, and the position, shape and form of these scars can be important in identification, particularly in winter, when other features are not available.

Veins. Leaves contain a network of harder tissue in the form of veins, which provide mechanical support for the leaf and carry the water- and food-material-bearing structures (the *xylem* and *phloem*, jointly), together with other tissues, forming the *vascular bundles*. In most leaves of Dicotyledons there is a prominent midrib which enters the leaf from the petiole and runs up the median line to the tip, giving off secondary branches which are themselves branched and ulti-

[13]

mately form a network of small veins. Venation of this type is known as *reticulate*. Another common type of venation is found mainly in Monocotyledons, and involves several, more or less equivalent, veins entering the leaf from the petiole or base, which run independently to the margins; these *parallel* veins are usually interconnected by smaller veinlets. Reticulate venation is essentially produced by a leaf which grows to its final size mainly around the margins, whereas parallel veins are produced in leaves which grow mainly at the base. In very thick or fleshy leaves it is usually not possible to see the venation, though holding the leaf up to the light can be helpful.

Ptyxis. This is the overall term to used to describe the manner in which young leaves are packed in the bud. It can be examined by sectioning a bud transversely, or by observing very young leaves as they emerge. This can be a helpful feature in attempting to identify shrubs in late winter or early spring. Most leaves are folded along the midrib, with the sides parallel and close together; this condition is described as *conduplicate*. In larger leaves, or leaves which are divided, each lobe or leaflet can be folded this way, producing a pleated effect, for which the term is *plicate*. Alternatively, the leaves of many species are rolled up into a tube in the bud, with one margin interior and the other exterior; this condition is known as *supervolute* and is familiar to people who grow the common house plants *Monstera pertusa* or *Ficus elastica*). In some other plants, e.g. violets (*Viola* species), the leaves have the margins rolled upwards and inwards in bud, a condition known as *involute*, whereas species of *Polygonum* have leaves whose margins are rolled in a reverse direction (downwards and inwards); this condition is known as *revolute*. Some leaves are flat or lightly curved, while others have combinations of the characteristics described above. In the insectivorous plants in the family *Droseraceae* the leaves are elongate and rolled from the tip to the base, either with the upper surface inside the spiral, or with it outside; in both these cases the ptyxis is described as *circinate*; (there are also a few rare cases not described here; for further information see Cullen, J., A preliminary survey of ptyxis (vernation) in the Angiosperms, *Notes from the Royal Botanic Garden Edinburgh* **37**: 161–214, 1978).

There are various other terms used to describe more minor characteristics of leaves; these are treated below under 'Miscellaneous features' (see p. 44).

3 Flowers

The flowers and the fruits that they lead to are the most important parts of the plant from the point of view of identification. They are also, of course, the most important parts from the point of view of the attractiveness of the plant. However, this is all problematic, because for long periods of the year, any individual plant will have neither flowers nor fruits; or, if it has flowers, it is possible that the fruits have not yet developed, or, again, if it has fruits it is possible that the flowers are already over. For the really accurate identification of many plants, both flowers and fruits are needed; both may be available at the later stages of flowering, which is therefore the best time to attempt identification, or, if the plant is growing in a garden, it can be examined twice, once in flower and once in fruit. In some cases (e.g. with many species of *Cotoneaster*) it is necessary to press the flowering specimens so that these can then be available when the plants have mature fruit, which is generally much later in the year.

A major problem consists in the definition of what is a flower. The word, as used in normal speech, is imprecise: the 'flower' of a dandelion is, in fact, an inflorescence (a collection of flowers forming a coherent whole). Flowers vary greatly from species to species, and it is difficult to find a definition that covers all the cases which occur. Possibly the best that can be done is to say that a flower is usually borne at the top of a long or short stalk (the *pedicel*) or, if stalkless, has its insertion on to some other organ and contains either one or more female sexual organs (*carpels*, see below) or one or more male sexual organs (*stamens*) or one or more carpels together with one or more stamens; there may be other parts associated with these as well, which may be protective in function (the *calyx*, composed of *sepals* which are free from each other or united at the base into a tube), or pollinator-attractive in function (the *corolla*, composed either of individual *petals*, which may be free from each other or united into a tube at the base). Nectar-secreting organs may also be present, formed by one or several *nectaries*, together with organs which are apparently reduced leaves (*bracts*, *bracteoles*, see below). Any of these organ-groups may be absent, depending on the particular type of flower. For instance, the female flower of a spurge (*Euphorbia*) consists of three united carpels (the ovary) only, whereas the flower of a catchfly (*Silene*) has bracts, bracteoles, sepals, petals stamens, a nectary and an ovary.

The number of possible combinations is very great, and the simplest way of discovering what is a flower in any particular plant is to find an ovary (or, if an ovary is entirely absent because the flowers are unisexual and male, a group of stamens which are obviously associated with each other, and to look at the organs immediately surrounding that. All these organ-groups, except for bracts and bracteoles, tend to occur in whorls of 2 or more, or, in rare cases, in compressed spirals. Even expert botanists are sometimes confused as to what exactly is a flower when confronted with a plant they have never seen before.

Inflorescences. Flowers are usually grouped together towards the ends of the branches of a plant into units which are called *inflorescences*. Again there is a problem of definition, because inflorescences are extremely variable. In general terms, an inflorescence may be defined as the arrangement of all the flowers on one branch, and it is often easier to see an inflorescence than to define exactly what it is. In most plants, as mentioned above, the inflorescences develop at the ends of young branches, but in some tropical plants, such as cocoa (*Theobroma*), the flowers are borne on the older, woody branches; this phenomenon is known as *cauliflory*. An inflorescence may consist of only a single flower (as in most tulips), or may contain many thousands of flowers.

In an inflorescence, each flower usually has its own long or short stalk (the pedicel), which attaches it to the axis of the inflorescence (known as the rachis). The stalk which bears the whole inflorescence (from above the uppermost leaf) is known as the *peduncle*. In plants with a single flower in the inflorescence, there is, of course, no easy distinction between pedicel and peduncle. There are many herbaceous plants in which the leaves are all borne in a rosette at ground-level, and the inflorescence is borne on a long or short stalk above them. Such a stalk is generally called a stem in common usage, but botanically it is a scape. Scapes are seen in dandelions (*Taraxacum* species) and in many species of *Primula*, to cite only familiar examples.

The pedicel of a flower is often borne in the axil of a small, leaf-like organ, known as a bract; in some cases (e.g. *Veronica*) there is little or no distinction between the upper foliage leaves and the lowermost bracts. On the pedicel there may be one, two or even more organs looking like the bracts but usually even smaller; these are known as bracteoles, and may bear further pedicels in their axils. Some plants, e.g. heathers (species of *Erica*) have both bracts and bracteoles, whereas others have bracts but no bracteoles (e.g. *Acanthus*), and yet others have neither (e.g. stocks, *Matthiola*).

The simplest inflorescence to appreciate is the *solitary* flower; this is seen, for example, in florist's tulips, where each stem terminates in a single flower. Similarly, scapes may also bear solitary flowers. Though the solitary flower is easy to understand, it is thought to have arisen, during the course of evolution, from various different types of inflorescence.

Aside from the solitary flower, there are two main kinds of inflorescence. In the first kind, the axis of the inflorescence continues to grow for some time, producing flowers as it grows, so that the oldest flowers are at the base of the inflorescence if it is elongate, or towards the outside if the axis is condensed. This type of inflorescence is a *raceme* (adjective: *racemose*). Again, there are basically two kinds of racemose inflorescence: if the flowers are stalked, the inflorescence is a raceme, whereas if they are stalkless (*sessile*), the inflorescence is a *spike*.

In the other main type of inflorescence, the apex of the stem ceases growth and becomes a terminal flower. Subsequent flowers arise on side-branches (generally borne in the axils of bracts) which originate from below the oldest flower (the terminal one). Hence, the oldest flowers are found towards the apex of the inflorescence if it is elongate or towards the centre if it is condensed. Such inflorescences are known

[17]

as *cymes* (adjective *cymose*), and there are many different types, which are produced from the simplest type by the suppression of particular branches.

Some plants bear compound inflorescences, either a raceme of racemes or a raceme of cymes. Such inflorescences are generally known as *panicles* (this term is sometimes restricted to the description of a raceme of cymes).

In some plants, particularly those belonging to the large daisy family (*Compositae*), the individual flowers are quite small and are aggregated together into inflorescences which can look like individual flowers. These inflorescences are known as *capitula* or *heads*, and are basically compressed racemes, as is shown by the fact that the outermost flowers mature earlier than the innermost. The bracts of the flowers form a calyx-like structure below the flowers, and this is known as an *involucre*. The individual flowers in each capitulum are generally stalkless, attached directly to the flattened or conical apex of the branch (the *receptacle*), and may be all of the same form, or the marginal flowers may differ from the central ones.

A further kind of complex inflorescence is the *catkin*, as found in many kinds of trees (*Alnus*, *Betula*, *Quercus*, etc.). They are complicated structures usually involving racemosely arranged bracts and/or bracteoles and flowers without corollas. Catkins may be upright or pendulous, and frequently fall as wholes, rather than breaking up and falling as individual flowers.

In a small number of cases, the real nature of the inflorescence is difficult to ascertain, even with detailed morphological studies. Such inflorescences are referred to as *clusters*, or *fascicles*.

In considering the individual flowers, it is best to deal with the various organ-groups separately, and they are discussed here from the centre of the flower outwards, as this provides the most sensible framework for explanation. When identifying a plant, of course, the flower is generally considered from the outside inwards, the various parts being recognised and counted before the flower is further dissected or sectioned (see below, p. 23).

The ovary. The ovary is present in all flowers unless the flower in question is purely male. It is found at the centre of the flower, and forms the uppermost (latest) part in terms of its origin (that is, it is the last part of the flower to be fully formed, and therefore is found at the morphological apex apex of the flower – this does not mean that it is necessarily the uppermost part of the flower when viewed, as later developments can take other parts of the flower physically above it). The ovary consists of, or is built up of, individual units known as carpels. There may be just a single carpel forming the ovary, or there may be 2 to many; if there are more than one, then they may be free (separate) from each other, or joined together (united, connate) into a compound structure. If the carpels are free from each other, the ovary is described as *apocarpous*; if the carpels are united to each other, then the ovary is described as *syncarpous*. The term '*gynoecium*' or even '*gynaecium*' is sometimes used as an alternative to 'ovary'.

The simplest case to consider (not necessarily the simplest case from the evolutionary point of view) is, of course, the ovary formed from a single carpel – the *unicarpellate* ovary. The most familiar example is that of the pea pod, as bought in a greengrocer's shop. This consists of the ovary of the pea flower as it develops into a fruit. It consists of an elongate bag, attached to the flower (and still bearing the calyx at the base) at one end, and tapering into the narrow style-base (see below) at the other end. The bag is the main body of the single carpel, and it contains the *ovules*, which later, when fertilised and fully developed, become the *seeds*. The ovules hang down on short stalks (*funicles*) from the upper angle of the bag, and are in two rows, though this can only be easily seen when the ovary is young. The tapering end of the pod, when it was younger, tapered to a thread-like portion, tipped by a velvety apex. The thread-like part is known as the *style* and the velvety apex is known as the *stigma*. In order for fertilisation of the ovules to take place, causing them to develop into the seeds, pollen of the correct type has to find its way to the stigma, where it begins to grow, as a fine tube, down through the style to reach the ovules. In peas bought from the greengrocer, the styles and stigmas have fulfilled their function and have been shed.

In hellebores (species of *Helleborus*) and winter aconites (species and hybrids of *Eranthis*), the ovary is made up of several to many carpels, each of which is similar to that of the pea. The carpels are

[19]

packed closely together in a whorl, but they are not at all joined to each other. In buttercups (species of *Ranunculus*), there are similarly many free carpels, but each one is much shorter, and contains a single ovule only. Such ovaries are described as *pluricarpellate* and apocarpous. Pluricarpellate apocarpous ovaries may have 2 to many carpels, and each carpel may contain 1 to many ovules. Each carpel in ovaries of this type has its own style and stigma, or, in some cases, the style is reduced, so that the stigma is borne directly on the body of the carpel (i.e. the stigma is sessile).

Unicarpellate or apocarpous ovaries are not all that common in the flowering plants; most have pluricarpellate ovaries in which the carpels are joined together into a compound structure in which the boundaries of the individual carpels are blurred or merged. Such compound ovaries are described as pluricarpellate and syncarpous. Syncarpous ovaries show varying degrees of joining of the carpels: the carpels themselves may be completely joined but the styles may be free from each other; or the carpels and styles may be joined to each other but the stigmas free; or all three may be fully joined; rarely, the carpels themselves may be almost entirely separate but the styles, at least, are joined together. All of these types including the last, which is found, for example, in *Asclepias*, are referred to as 'syncarpous'.

In joining together, the individual carpels may retain their individual internal spaces, or these internal spaces may merge into a single internal space. Thus a syncarpous ovary may contain several ovule-containing spaces which are separated by cross-walls, or there may be a single internal space which contains ovules contributed by the various carpels. The ovule-containing spaces are known as *loculi* or *cells*. An ovary containing several loculi is known as *multilocular* (or 2-, 3-, 4- or more-locular), whereas an ovary containing only a single space is known as *unilocular* (the word '*celled*' may be substituted for 'locular' in the above terms). An ovary may be pluricarpellate and unilocular and contain only a single seed, contributed by one of the carpels.

It is important for accurate identification to know the number of carpels which make up a pluricarpellate ovary. The best way of deciding this is to count the number of stigmas, if they are free from each other; this is generally the same as the number of carpels. If the stigmas are united, then counting the number of loculi in the ovary (if there are more than one) will usually provide the same information (in a few families, particularly *Linaceae*, the loculi become further divided by ingrowing walls, so that the number of loculi is twice the number of carpels). If the stigmas are united and the ovary is unilocular, then it is very difficult to decide how many carpels there are (microscopic study may be required). In such cases, however, experience shows that the number of carpels is usually either 2, or the same as the number of petals/corolla-lobes.

In order to see how many cells the ovary contains, it is usually necessary to section it horizontally across the middle. Such a section will also generally show how many ovules there are in the ovary or in each loculus, and, very often, the way they are attached to the ovary – the placentation, which is a very important characteristic in the identification of the families (see below). With some groups it may also be necessary to cut another ovary longitudinally (along a diameter) to see precisely how the ovules are attached.

The cutting of the ovary, as described above, is an art that has to be mastered with practice, as the ovaries can be very small indeed. In general terms it is relatively easy to cut a transverse section, using a single-edged razor blade. The section should usually be taken at the widest part of the ovary, and both the cut ends should be looked at. If necessary, further thin sections should be cut from the cut ends, so that the structure can be more clearly seen. Longitudinal sections can be difficult and troublesome. The flower (or just the ovary, if it is large enough) should be held between finger and thumb, with the flower in the same position as it was on the plant, with the stem end pointing outwards. A cut should then be made carefully through this, moving the blade downwards as the cut goes through the tissues. In this way, clean surfaces will be left and the ovary will not be squeezed or squashed, as will happen if a sawing motion of the blade is used.

It is sometimes helpful to rub over the ends of cut sections with the tip of a broad, black, felt-tip pen. The water-based ink is absorbed by the cut surfaces, which therefore stain black, but is repelled by the waxy surface of the ovules, which are pale green and stand out from the rest of the stained cut.

Placentation. This is a very important character in the identification of the families, and should ideally be observed in both transverse and longitudinal sections of the ovary (as above). Though there is some overlap between the different types of placentation, it can be conveniently dealt with under the terms used in the key.

(i) *Marginal.* This term is used only in those cases in which the carpels are free (or the ovary is made up of a solitary carpel) and describes the condition in which the carpel bears several ovules on its upper suture (e.g. *Caltha, Pisum,* etc., see fig. 1*a, b*). If there is only a single ovule in the carpel, it may be borne at the base, when it is described as [*marginal*]- *basal,* or at the apex, when described as [*marginal*]-*apical* (see fig. 3*h, k*).

(ii) *Axile.* In this condition the ovary is always made up of 2 or more fused carpels and contains cross-walls (*septa*) which form the loculi or cells. The ovules are borne on the central axis, where the cross-walls meet (e.g. *Narcissus,* see fig. 1*g*), on swollen placentas (e.g. *Solanum,* see fig. 1*c–f*) or on intrusive placental outgrowths (e.g. some species of *Begonia* (see fig. 1*i*). In some families the ovules are reduced to 1 or 2 in each cell and ascend from the base (e.g. *Ipomoea,* see fig. 1*l, m*) or are pendulous from the apex (e.g. many Umbelliferae, see fig. 1*b.* Ovules in axile ovaries sometimes occur side-by-side (*collateral* ovules) as in *Heliotropium* (see fig. 1*l, m*), or one above the other (*superposed* ovules) as in most Acanthaceae (see fig. 1*j, k.* Occasionally axile ovaries are further divided by secondary septa, which grow inwards from the carpel wall as the ovary matures (e.g. *Linum, Salvia*), so that the ovary comes to have twice as many cells as carpels.

(iii) *Parietal.* This term is used when the ovules are borne on the walls of the ovary, or on outgrowths from them. Several situations may be distinguished.

In the majority of cases, parietal placentation occurs in 1-celled (unilocular) ovaries made up of several united carpels, the ovules being restricted to placental regions on the walls, as in *Viola* (see fig. 2*a*), *Gentiana* (see fig. 2*d–f*) or *Ribes,* or on intrusive, placenta-

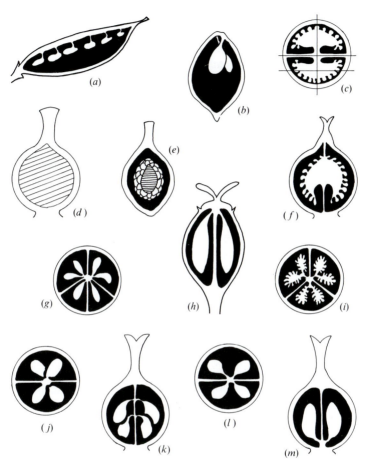

Figure 1. Placentation (see pp. 22–27). (*a*), (*b*) marginal; (*c*)–(*m*), axile. (*c*)–(*f*) ovules on swollen placentas ((*c*) transverse, (*d*)–(*f*) longitudinal sections; planes of the longitudinal sections indicated in (*c*)). (*g*) ovules borne on the axis, (*h*) ovules pendulous. (*i*) ovules on intrusive placentas. (*j*), (*k*) ovules superposed. (*l*), (*m*) side-by-side.

bearing outgrowths from them (e.g. *Cistus*, *Heuchera*, see fig. 2*b*). Intrusive parietal placentas may almost meet in the middle of the ovary, so that the distinction between axile and parietal placentation is not always clear-cut (e.g. *Escallonia*, *Cucumis*, see fig. 2*g–i*). Gently squeezing the transversely cut ovary will show whether or not the placentas are united in the centre of the ovary.

In a few cases the ovules are borne on the walls of a 2- or more-celled ovary; this is found particularly in the *Aizoaceae*. In most *Cruciferae* the older ovary and fruit are 2-celled, but this is because of the development, during ripening of the fruit, of a false septum (*replum*) across the ovary (see fig. 2*c*).

Occasionally the ovules are scattered over most of the inner surfaces of the carpels. This situation is distinguished as *diffuse-parietal* placentation; it can occur in ovaries with free carpels (e.g. *Butomaceae*, see fig. 2*j*) or of united carpels (e.g. *Hydrocharitaceae* (see fig. 2*k*).

As shown in the accompanying diagrams (figs. 1, 2 & 3), the side view of axile and parietal placentation can vary greatly according to the vertical plane in which the ovary has been cut. The longitudinal section can be best understood in relation to a transverse section.

(iv) *Free-central*. In this condition, the ovules (usually many) are borne on a central spherical or columnar structure that rises from the base of a 1-celled ovary made up of several united carpels (e.g. *Pinguicula*, see fig. 3*a–c*). In most cases, a thread of tissue attaches this placental column to the top of the ovary; sometimes this thread is rather stout (e.g. *Lysimachia*, see fig. 3*c*). Occasionally the ovary may be septate near the base (e.g. *Silene*) although in most ovaries with free-central placentation, such septa break down as the ovary matures.

(v) *Basal*. Here the ovules arise from the base of a 1-celled ovary (or rarely from the base of a solitary or free carpel, see above), as in *Polygonum*, *Tamarix*, *Armeria*, etc. (see fig. 7 (*d*), (*e*), (*g*)) or are borne on a basal placental cushion (oblique in *Berberis*, see fig. 3*f*).

[24]

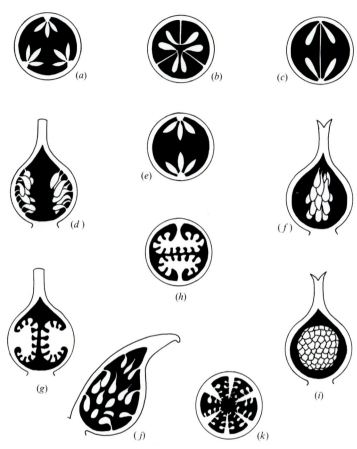

Figure 2. Placentation (see pp. 22–27). Parietal types (a) ovules on the carpel walls, (b) ovules on intrusive placentas, (c) ovules on the carpel walls, septum present. (d)–(f) ovules on carpel walls: (d) longitudinal section through placentas, (e) transverse section, (f) longitudinal section at right angles to placentas, (g)–(i) ovules on intrusive placentas which almost meet in the centre of the ovary; (g) longitudinal section through the placentas, (b)transverse section, (i) longitudinal section between the placentas, (j), (k) diffuse parietal.

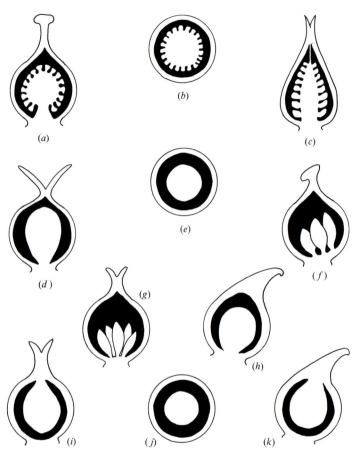

Figure 3. Placentation types (see pp. 22–27). (*a*)–(*c*) free-central; (*d*)–(*h*) basal; (*i*)–(*k*) apical, (*a*), (*b*) ovules free-central, (*c*) ovules free-central, showing attachment of placenta to top of ovary, (*d*), (*e*) one basal ovule, (*f*) ovules on an oblique placental cushion, (*g*) several basal ovules (ovary of united carpels). (*h*) one basal ovule (ovary of free carpels), (*i*), (*j*) ovule apical (ovary of united carpels, (*k*) ovule apical (ovary of free carpels).

(vi) *Apical*. In this case the ovule (generally solitary) is attached to the apex of the single cell (or free or solitary carpel, see above), as in *Scabiosa* (see fig. 3*i*, *j*), or *Anemone* (see fig. 3*k*).

Although it would be possible to describe the ovules in the cells of many septate ovaries as 'apical' or 'basal', we have not used the terms to cover these situations. Instead, to avoid confusion, 'pendulous' or 'ascending' have been used, and are considered to be special cases of axile placentation.

The stamens. The male parts of the flower are called stamens, collectively forming the androecium. They are generally found in 1 or more whorls outside the ovary in bisexual flowers, but they may be apparently central in the purely male flowers of some unisexual species. Each stamen is a relatively simple structure, consisting of a stalk, the *filament*, which is usually thread-like, but can be quite thick, bearing at the top the *anthers* (the pollen-containing parts). These are usually broader than the filament and borne at its apex, but can be narrow and somewhat sunk in the broad filament. The anther is usually made up of 4 sacs (which may become confluent into 2) which contain the pollen. The pollen is usually dry and granular, but occasionally it is sticky (e.g. species of *Rhododendron*) and sometimes aggregated into masses, known as *pollinia*, which are dispersed as wholes (found in orchids, species of *Asclepias* and its relatives, *Mimosa*, etc.). In a few groups (e.g. most *Ericaceae*, *Droseraceae*, *Juncaceae*, etc.) the pollen grains are not separate, but are aggregated in groups of 4 (*tetrads*); this is difficult to see without the aid of a compound microscope, as are the surface features of the pollen grains themselves, which can be very important for identification; these features are, however, beyond the scope of the present book.

The number of stamens, as well as their structure, is important in identification. There may be only a single stamen per flower (e.g. *Euphorbia*), or there may be 2 to many. In some plants the stamens in each flower are joined together (united, connate), sometimes just by their filaments (as in many plants related to the mallows, *Malva*), or by their anthers (as in the family *Compositae*), or by both (e.g. some members of the family *Meliaceae*); when joined by their fila-

[27]

ments, the tube so formed generally surrounds the ovary and style.

In order for the pollen to be released, the anthers must open. The most frequent method is for slits to develop down the length of the pollen-sacs; there are generally 2 slits, one on either side. In a few groups the anthers open by distinct pores either really at the top (*Polygalaceae*) or apparently so (*Ericaceae*; during development of these stamens, the anther becomes inverted, so that the pores which are apparently at the top are, in fact, at the morphological base). In a small number of cases (*Berberis*, *Hamamelis*, etc.) the anthers open by flaps or valves which may curve upwards or outwards. The opening of the anthers is referred to as dehiscence: the normal method is known as *longitudinal* dehiscence, that by pores as *porose* (or *poricidal*) dehiscence, and that by valves as *valvular* dehiscence. Porose dehiscence is found in the following families: *Ochnaceae*, *Elaeocarpaceae*, *Tremandraceae*, *Polygalaceae*, *Melastomataceae*, *Actinidiaceae* (part), *Ericaceae* (most) and *Mayacaceae*, but occurs sporadically in other groups (e.g. the snowdrop, *Galanthus*, in the *Amaryllidaceae*, *Cassia* in the *Leguminosae*).

In some functionally female flowers, and also in some bisexual flowers, sterile, stamen-like structures may be found either where the genuine stamens might be expected, mixed with the genuine stamens, or outside of them. These are known as *staminodes*. In bisexual flowers, the origin of the staminodes from the stamens is generally fairly obvious (e.g. *Parnassia*, *Sparrmannia*) though occasionally staminodes may be difficult to distinguish from the petals (e.g. *Aizoaceae*).

The perianth. The ovary and stamens form the most important parts of the flower, and indeed many flowers consist of these organs only, either singly or together. However, in most flowers there are other protective and/or pollinator-attracting structures to the outside of the stamens; these are collectively known as the *perianth*, and may consist of distinct calyx and corolla, or of a single series of organs, or, more rarely, of several series of organs. The organs of the perianth are generally arranged in whorls, though in a few families (e.g. *Magnoliaceae*) they are arranged in compressed spirals. A perianth with a single whorl of organs is known as *uniseriate*, one with 2 whorls as *biseriate*; and one with 3 or more whorls as *multiseriate*.

The most frequently seen case is the biseriate perianth, consisting of two whorls of organs. Those of the outer whorl are known collectively as the *calyx*, whose individual organs are known as *sepals*, and those of the inner series as the *corolla*, whose individual organs are known as petals. If the sepals are joined to each other at the base to form a tube, then the individual parts are referred to as *calyx-lobes* and the calyx is referred to as *gamosepalous* (as opposed to *polysepalous* for the state in which the sepals are free); similarly, if the petals are joined together into a tube, then they may be referred to more precisely as *corolla-lobes* and the corolla as *gamopetalous* (as opposed to *polypetalous*). It is sometimes difficult to decide whether or not the individual organs of the whorl are joined to each other (*connate*) or not. Generally, a corolla of united lobes falls as a whole, whereas one with free petals falls as free petals.

In general terms, the sepals are usually greenish and somewhat leaf-like, and serve to protect the delicate interior organs of the flower in bud. Usually, the sepals have 3 veins entering them from the base; this, is often difficult to see, either because of the thickness of the sepal itself, or because the 2 outer veins are often much less pronounced than the central one, so that the sepals appears to be 1-veined. The petals, on the other hand, are usually thinner, larger and more brightly coloured than the sepals, usually serving to attract pollinators to the flower. There are, however, many exceptions to these statements. The petals are generally 1-veined from the base, and this can usually be relatively easily seen.

Flowers in which the perianth consists of a single whorl are generally considered to have no petals, the perianth consisting of sepals only. A more neutral terminology in these cases is to describe the organs of the whorl as *perianth-segments* or *tepals*. This is so even if these organs are brightly coloured and petal-like (as in many species of *Clematis*). However, flowers which appear superficially 1-whorled, may have 2 whorls, the calyx being extremely small and reduced (as in many species of *Rhododendron*); this must be looked for very carefully. Some further guidance on deciding whether sepals

[29]

and petals or perianth-segments are present is given below under 'horizontal arrangement of parts' (p. 31).

In the flowers of some species there is no perianth at all. Such flowers are usually wind-pollinated, and the sexual parts are often associated with bracts and/or bracteoles (e.g. birch, beech and oak, which bear their flowers in catkins, and grasses and sedges).

The symmetry of the perianth is often an important character in identification, and it is one that often causes trouble in interpretation, because nature is never totally symmetrical. Many flowers are built on a radial plan, so that the perianth forms more or less a circle when viewed in outline. Such perianths have many planes of symmetry (sectioning the flower along any radius will produce 2 mirror-image halves), and are described as *actinomorphic* or *radially symmetric*. Many common flowers such as the rose and buttercup have flowers of this type. Others have a perianth with only a single plane of symmetry; such flowers appear to be 2-sided when viewed in outline, and are described as *zygomorphic* or *bilaterally symmetric*. Usually, the single plane of symmetry is the vertical, but occasionally it is horizontal (as in *Corydalis*). In a few plants (species of *Maranta*), the perianth is asymmetric, but the flowers are borne in pairs, so that the combined perianths of the pair of flowers are zygomorphic.

As well as the perianth, stamens and ovary can be zygomorphically arranged. This generally occurs when the perianth is also zygomorphic, though it also occasionally occurs where the perianth is actinomorphic. In many *Ericaceae*, for instance, the 10 stamens are deflexed downwards (*declinate*) in a group, arching towards the lower part of the flower and then arching upwards again towards the corolla opening; and in *Gloriosa* (a tropical climber in the *Liliaceae*), the style is borne at right angles to the axis of the flower.

Nectaries. In some flowers (e.g. *Magnolia*, *Papaver*), no nectar is produced, but nectar is found in most. It may be sticky and in small quantity, or more watery and copious, and is secreted by zones or rings of tissue called *nectaries*. These may be found on the perianth (on the inner surfaces of the sepals, as in many *Malvaceae*; on the petals, as in many *Ranunculaceae*), or on the floral receptacle, either between the petals and stamens or the stamens and ovary, or both, or on the ovary itself (many *Liliaceae*). Nectar produced by the nectaries is sometimes held in *spurs* (generally backwardly-projecting, narrow sacs) either of one or more of the sepals or petals.

Horizontal disposition of parts of the flower. Flowers are immensely variable in the way their parts are arranged; in this section the horizontal arrangement of the parts is described, while the vertical arrangement is treated in the next section.

It is simplest to consider first radially symmetric flowers, whose outline forms more or less a circle, in which the stamens are twice as many, or as many as the sepals or petals. As an example, a flower with parts in 5s will be described, but the principles apply to flowers with different numbers of parts.

In such a flower, the 5 sepals will be arranged symmetrically, their apices forming part of a circle, and imaginary lines down their centre will lie at approximately 72° from each other. If we now look at the 5 petals, we will find that they are disposed in the same way, but alternating with the sepals, so that the imaginary mid-line along each petal lies exactly in between those of 2 of the sepals. Then, if we look at the stamens, various conditions arise. Generally, if there are 5 stamens, they will usually be found on the same radii as the sepals; in a few cases (e.g. *Primula*), this is not so, and the stamens are found on the same radii as the petals, when they are described as *antepetalous*. If there are 10 stamens, it is usual for them to be arranged in 2 whorls of 5, those of the outer whorl on the same radii as the sepals, those of the inner whorl on the same radii as the petals. In the rare case that the 5 outer stamens should be on the same radii as the petals and the 5 inner on the same radii as the sepals, the flower is described as *obdiplostemonous*; this is very unusual, but occurs in species of *Geranium*. The principle is that the organs of each whorl (whether they are free or fused) lie on alternating radii; it is thought that suppression of various whorls leads to the situations found in nature. Thus, if a flower has a uniseriate perianth of 5 segments, with the 5 stamens on the same radii as the perianth-segments, then it is likely that the latter are sepals (petals having been suppressed or lost during evolution). Similarly, if the perianth has 5 segments and there are 5 stamens on radii

[31]

alternating with those of the perianth-segments, then it is possible that either the segments are petals (sepals having been suppressed or lost during evolution), or that they are sepals and petals and a whorl of stamens has been suppressed or lost.

Antepetalous stamens are characteristic of the following families: *Berberidaceae, Lardizabalaceae, Sabiaceae, Rhamnaceae, Leeaceae, Vitaceae, Myrsinaceae, Primulaceae* and *Plumbaginaceae*.

The alternation of radii will continue with respect to the carpels of the ovary, if these are of the same number as the sepals, petals and stamens; otherwise it is not usually possible to determine which radii the carpels are actually on.

The same principles can now be applied to zygomorphic flowers, though with these, the shape of the perianth sometimes causes difficulties in determining the various radii. When this is the case, it is necessary to look at the base of each whorl of organs to see how they are arranged with respect to each other.

In bud, the sepals and petals (or perianth-segments) are often arranged in characteristic ways. This phenomenon is called *aestivation*, and is sometimes of importance in identification. In the simplest case, the organs lie edge-to-edge in bud, without overlapping (indeed, sometimes with quite considerable spaces between them). This is known as *valvate* aestivation (see fig. 4*d*). In other cases, the various segments of the whorl overlap each other; the general term for this kind of aestivation is *imbricate* (see fig. 4*a–c*). There are various kinds of imbricate aestivation, depending on the alignment of the various organs; the only one of significance for family identification is that in which each segment overlaps, and is overlapped by, one other. Such aestivation is known as *contorted* (fig. 4*c*), and can be seen, for example, in the periwinkle (species of *Vinca*).

Vertical disposition of the parts of the flower. Two sets of terms are used in describing the relative vertical positions of the attachment of the floral organs. One (superior/inferior) is used with reference to the position of the ovary with respect to the other floral organs. The other (hypogynous/perigynous/epigynous) refers to the position of the other floral organs with respect to the ovary, but is best

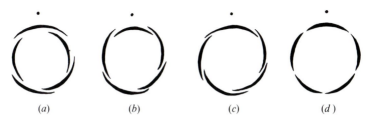

(a)   (b)   (c)   (d)

Figure 4. Aestivation types, (a), (b) overlapping (imbricate) — the details of the manner in which the organs overlap each other is variable and two of the common types are shown. (c) contorted (each overlapping one other and overlapped by one other — a special case of imbrication), (d) edge-to-edge (valvate).

explained in terms of the fusion of the organs of different floral whorls.

The terms referring to ovary position are not especially ambiguous: a *superior ovary* is one borne on the receptacle (the apex of the pedicel) above the insertion of the other floral organs (regardless of whether these are free from each other or variously united); an *inferior ovary* is one borne below the point of insertion of the other floral organs (free or united) so that they appear to be borne on top of, or at least on the upper parts of the sides of, the ovary. A rather rare intermediate condition, in which the lower part of the ovary is inferior and the upper part superior (i.e. the other floral whorls appear to be borne about halfway up the ovary) is referred to as *half-inferior* or *semi-inferior*.

The position of the ovary is best determined using a longitudinal section of the flower. However, it is generally possible to tell if the ovary is inferior by looking at the back of the flower, when, if it is inferior, it will be seen projecting below the base of the calyx.

The other terminology is more difficult to apply. One of the difficulties is that various authors have applied it to the whole of the flower. This practice is misleading, and the terms should be used to refer to the perianth and/or stamens only. The following table (p. 34) and the accompanying illustrations should make clear the proper usage of these terms.

Table 1. *Relationships of floral parts (see fig. 5)*

Ovary (G) position	Fig. 5	Insertion of perianth (P or K & C) and androecium (A)	Description adopted here	Description used in older literature
	(a)	PA or KCA inserted independently on receptable (e.g. *Ranunculus*)	PA or KCA hypogynous	Flower hypogynous
	(b)	K & C apparently fused at base, A inserted independently on receptacle (e.g. *Tropaeolum*)	K & C perigynous borne on a perigynous zone, A hypogynous	Various
Superior	(c)	C & A apparently fused at base, K inserted independently on receptable (e.g. *Primula*)	K hypogynous, C & A perigynous borne on a perigynous zone	Flower hypogynous, A epipetalous
	(d)	K, C & A inserted on a ring or collar of tissue which is inserted on receptable (e.g. *Prunus*)	K, C & A perigynous borne on a perigynous zone	Flower perigynous
	(e)	P & A apparently fused, C absent (*e.g. Daphne*)	P & A perigynous	Various

[34]

Table 1. *cont.*

Ovary (G) position	Fig. 5	Insertion of perianth (P or K & C) and androecium (A)	Description adopted here	Description used in older literature
	(f)	P & A or K, C & A inserted independently on walls of ovary (*e.g. Paliurus*, some species of *Saxifrage*)	P & A or K, C & A partly epigynous	Various
	(g)	P & A or K, C & A inserted independently apparently on top of ovary (*e.g. Umbelliferae*)	P & A or K, C & A epigynous	Flower epigynous
Fully inferior	(h)	K, C & A inserted on top of ovary, C & A fused (e.g. *Viburnum*)	K, C & A epigynous, C & A borne on an epigynous zone	Flower epigynous
	(i)	K, C & A inserted on a ring or collar of tissue itself inserted on top of ovary (e.g. *Fuchsia*)	K, C & A epigynous, C & A borne on an eiigynous zone	Flower epigynous

The 'ring' or 'collar' of tissue mentioned in the table, probably best referred to as a *perigynous zone* if the ovary is superior and as an *epigynous zone* if the ovary is inferior is often referred to as 'calyx', 'floral cup', 'floral tube' or 'hypanthium'.

[35]

These terms may be difficult to use with unisexual flowers. If a *pistillode* (a sterile ovary) is present in a male flower, it may be possible to decide whether the various whorls are hypogynous, perigynous or epigynous. If, however, no pistillode is present, much care must be taken, and a female flower must be sought. If the female flower completely lacks a perianth (e.g. in *Betula*) then it is not possible to use these terms at all, and such ovaries are described as *naked*.

The following situations occur in the families covered in the key.

I. Perianth and stamens hypogynous; ovary superior. See fig. 6*a*, *b*.
II. Calyx hypogynous; corolla and stamens perigynous; ovary superior. See fig. 5*c*, *d*.
III. Perianth (calyx and corolla) perigynous; stamens hypogynous; ovary superior. See fig. 7*e*.
IV. Perianth (calyx and corolla) and stamens perigynous; ovary superior. See fig. 7*a–c*.
V. Ovary partly or fully inferior; perianth (calyx and corolla) and stamens epigynous, without an epigynous zone. See fig. 7*d*, *e*.
VI. Ovary partly or fully inferior; perianth (calyx and corolla) and stamens with an epigynous zone. See fig. 8*a*, *b*.

Types III and IV are often complicated by the presence of a *nectariferous disc* (a disc or ring of nectar-secreting tissue) surrounding the ovary, which may sometimes seem to be immersed in it. Usually, when such a disc is present, the perianth (as a clearly recognisable structure) is inserted on the edge of it. The stamens may also be borne on the edge of the disc, as in many species of maple (*Acer*) or on top of the disc, as in the spindle tree (*Euonymus*). In the flowers of some families the ovary is borne on a long or short stalk (*gynophore*); this can be ignored when deciding whether the parts are hypogynous or perigynous (a gynophore is not possible when the parts are epigynous). A further complication arises in the passion flower (*Passiflora*) in which the ovary and stamens are borne on a common stalk (*androgynophore*) within the flowers; and in some members of the *Caryophyllaceae* (e.g. *Silene*), the corolla, stamens and ovary are borne on a short common stalk (known as an *anthophore*) within the flower.

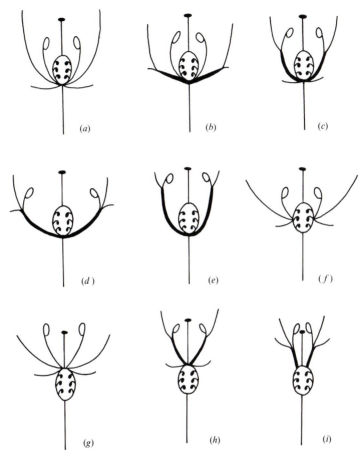

Figure 5. Diagrams illustrating the usage of the terms hypogyny, perigyny and epigyny. Perigynous and epigynous zones are indicated by the use of heavy lines. For further information see table 1.

The terminology described above (hypogynous, perigyny, epigyny) is traditionally applied to the flowers of the Dicotyledons; the superior/inferior terminology is applied to both Dicotyledons and Monocotyledons. In the descriptions of the families on pp. 91–187, the ovary position in the Dicotyledons can be assumed from

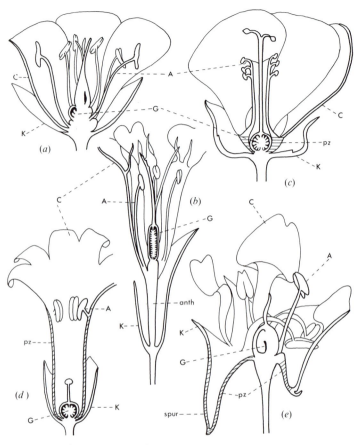

Figure 6. Relative positions of floral parts (see p. 36). Type I: (*a*) *Geranium*, (*b*) *Silene*; type II: (*c*) *Abutilon*, (*d*) *Primula*; type III: (*e*) *Tropaeolum*. A – androecium, anth – anthophore, C – corolla, G – gynoecium, K – calyx, pz – perigynous zone (shaded).

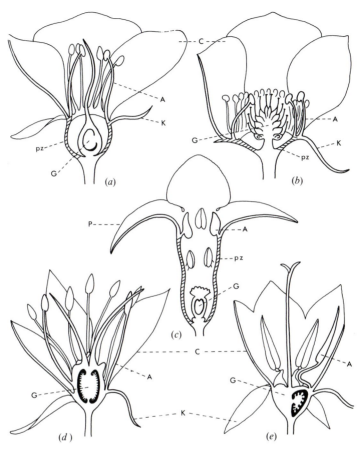

Figure 7. Relative positions of floral parts (see p. 36). Type IV: (*a*) *Prunus*, (*b*) *Geum*, (*c*) *Daphne*; type V; (*d*) *Saxifraga stolnifera*, (*e*) *Campanula*. A – androecium, C – corolla, G – gynoecium, K – calyx, P – perianth (undifferentiated), pz – perigynous zone (shaded).

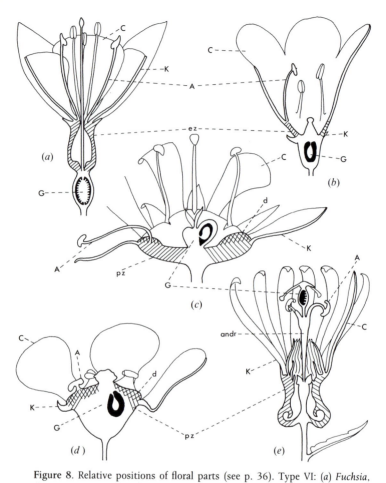

Figure 8. Relative positions of floral parts (see p. 36). Type VI: (a) *Fuchsia*, (b) *Viburnum*. More complicated types: (c) *Acer* (see p. 36), (d) *Eunonymus* (see p. 36), (e) *Passiflora* (see 36). A – androecium, andr – androgynophore, C – corolla, d – disc (cross-hatched), ez – epigynous zone (hatched), G – gynoecium, K – calyx, pz – perigynous zone (hatched).

the description of the other parts; in the Monocotyledons the ovary position is given for each family.

4   Fruits

In strict botanical terms, the fruit is the maturation-product of the ovary of a single flower. However, the term can be used more loosely to describe the structure which opens to release the seeds for dispersal, or falls or is removed from the parent plant for the same purpose. This structure may be simply the fruit as strictly defined above, or it may involve other parts of the flower (sepals, styles, bracts, etc.), when it is known strictly as a *false fruit*, or the coalesced ovaries of several flowers, when it is known strictly as a *compound* fruit. The looser sense of the term will be used throughout this section.

In maturation into the fruit, the ovary wall becomes the fruit wall or *pericarp*. In many fruits the pericarp has a tripartite structure, with a tough outer rind (the *exocarp*), a fleshy or fibrous central part (the *mesocarp*) and a hard or stony inner part (*endocarp*), which surrounds the individual seeds. The ovule(s) in the ovary mature into the individual seed(s), which are described in the next section.

There are many different kinds of fruit known in the flowering plants, and there is a very complex terminology used to describe them. Only those most frequently encountered will be described here.

The criteria defining the different types of fruit include those mentioned above and several others:

(a) whether the ovary from which the fruit was formed was apocarpous or syncarpous, and whether it was superior or inferior;
(b) whether the fruit has a defined opening mechanism (a *dehiscent* fruit) or not (an *indehiscent* fruit);
(c) whether the fruit is fleshy or not, the fleshiness contributing to its dispersal by animals;
(d) how many seeds the fruit contains (1, 2 or more).

Follicle. A fruit derived from a single free carpel. Hence the actual fruit may be a single follicle (if the ovary from which it is derived was of a single carpel, e.g. *Consolida*) or a group of follicles (if the ovary was apocarpous and pluricarpellate, e.g. *Helleborus*). The follicle is dry, contains several seeds and dehisces by opening out along its inner suture.

Legume. The legume or *pod* is similar to the follicle, but opens along both sutures. It is the characteristic fruit of the family *Leguminosae*.

Achene. A 1-seeded, dry, indehiscent fruit which may be formed from an apocarpous or syncarpous, superior ovary (when formed from an apocarpous ovary the fruit is strictly a group of achenes). Achenes are formed in many families.

Cypsela. The equivalent of the achene when formed from an inferior ovary (e.g. all *Compositae*).

Caryopsis. Essentially an achene in which the pericarp and the seed coat (testa) have fused together, characteristic of the grass family (*Gramineae*).

Capsule. Formed from a pluricarpellate ovary, the capsule is dry, several-seeded and dehisces by longitudinal (or rarely radial) splitting of the pericarp or by the formation of pores within the pericarp. It is the most common type of fruit.

Lomentum. The equivalent of a follicle, legume or capsule which breaks up into 1-seeded portions which are themselves indehiscent. This type of fruit is found in the *Leguminosae* and *Cruciferae* and occasionally in other families.

Schizocarp. An indehiscent, dry fruit which splits into 2 or more, generally 1-seeded parts (*mericarps*), which usually represent the individual carpels from which the ovary was formed (e.g. *Umbelliferae*).

Samara. An achene, cypsela or mericarp which develops a conspicuous wing which aids in its dispersal (e.g. *Acer*).

[42]

Nut. A large achene or mericarp with a very hard pericarp. Smaller examples (e.g. *Polygonum*) are referred to as 'nutlets'.

Berry. Essentially an indehiscent capsule in which the pericarp is fleshy and succulent. There may be 1 or more seeds (e.g. *Ribes*) or, if the endocarp becomes hard and bony around each seed, there may be 1 to several *stones* or *pyrenes*.

Drupe. An indehiscent fruit, usually containing a single seed, in which the pericarp is clearly tripartite, with tough exocarp, fleshy mesocarp and hard endocarp. These are found especially in species of *Prunus* (cherry, plum, etc.). In *Rubus*, which has an apocarpous ovary, the fruit consists of a group of united, small drupes, known as *drupelets* (raspberry, blackberry).

Pome. A fruit found only in those members of the *Rosaceae* which have inferior ovaries. Really a false fruit, involving the ovary and fleshy receptacular tissue which forms around it. The ovary inside the pome may be had and stone-like (as in *Cotoneaster*, *Crataegus*, etc.) or parchment-like (*Malus*, *Pyrus*).

There are many other rare types of fruit, but these are generally not found in the families included in this book.

5 Seeds

The seeds are the maturation-products of the ovules, and contain the potential for the next generation. They are generally dispersed from the parent plant within the fruit, but may be dispersed individually. Each seed contains an embryo (which may be rudimentary at dispersal, as is the case in orchids and many parasitic plants) and possibly food-reserve material (*endosperm* or *perisperm*), wrapped inside the seed-coat or *testa*.

Seeds vary in size from minute and dust-like (as in many orchids) to large and solid. The testa may be variously marked and coloured and the seed may be appendaged. Two kinds of appendage are important: the *aril*, which is an outgrowth from the funicle (the stalk of the ovule) and is often fleshy or coloured and may partly

[43]

or wholly envelop the seed, and the *elaiosome* (also known as the *caruncle*), which is an oily body formed at one end of the seed.

6 Miscellaneous features

Many plant parts have an *indumentum*, that is, they are covered with hairs of various forms. Hairs may be simple (unbranched), branched, bifid (with 2 arms, attached in the middle) or stellate, or modified into scale-like structures (*peltate scales*). Simple hairs may be unicellular or multicellular (difficult to see without the aid of a microscope), and unbranched hairs may terminate in a glandular tip.

There are many different terms to describe the hair covering, some of which (those used in the keys) are treated in the Glossary (pp. 195–205).

Most plants have clear sap, but in some families the sap is milky and/or coloured. This feature should be easily seen when a leaf is broken from a stem, but sometimes the exudation of the sap is slow, so time must be taken. Care should also be taken in dealing with plant-sap, which is often irritant or poisonous.

Plants frequently bear *spines* on their stems or leaves; these are generally outgrowths of the stem, and serve for protection from browsing animals, or as hooks for climbing or scrambling over other plants. Many climbers bear *tendrils*, fine, thread-like structures which coil around other vegetation or supports, enabling the plant to climb. Tendrils may be modified leaves, modified leaflets or modified inflorescences. In some cases the tendrils end in sticky pads which enable the plant to grip the support more closely. In a few climbers (e.g. ivy, *Hedera*), specialised climbing roots grip the supports.

Very specialised leaves are found in insectivorous plants, which may be modified into pitchers or active insect-traps. Botanical textbooks should be consulted for further information on these organs.

In some other families, the leaves contain glands or ducts full of aromatic oils. These can usually be seen as translucent dots or lines when the leaf is held up to the light, and the oils can be smelt when the leaf itself is crushed (e.g. *Eucalyptus*).

Using the keys

The keys in this book are of the bracketed type, and are dichotomous throughout, i.e. at every stage a choice must be made between two (and only two) contrasting alternatives (leads), which together make up a couplet. As the main key allows for the identification of 258 families, it has been arranged in groups, with a key to the groups at the beginning. To facilitate reference to particular leads, each couplet is numbered, and each lead is given a distinguishing letter (a or b).

To find the family to which a specimen belongs, one starts with the key to the groups and compares the specimen with the two leads of the couplet numbered 1. If the specimen agrees with 1a, one proceeds to the lead with the number that is the same as that appearing at the right hand end of 1a (in this case, 2); if, however, the plant agrees with 1b, then one proceeds to the couplet numbered 12. This process is repeated for subsequent couplets until, instead of a number at the right hand end of a lead, a group is reached. Throughout this process, it is very important that the whole of each couplet is carefully read and understood before making a decision as to which lead to follow.

One proceeds in the same way within the group keys until the name of a family is reached. The families are numbered in the key, and, to provide a check on the identification obtained, the families are briefly described, in numerical order, on pp. 91–187. The specimen should be compared carefully with the description of the family; this should help to reveal errors made in observation and the use of the key.

In order that back-tracking should be easy, the number of the lead from which any particular couplet is derived is given in brackets after the couplet number of the 'a' lead. Thus '12a. (3)' means that one arrived at couplet 12 from one of the leads of couplet 3.

It will sometimes happen that the specimen does not agree with all the characters give in a particular lead. When this situation arises, one must decide which of the two leads of the relevant couplet the specimen agrees with most fully. In general, the most reliable diagnostics are put at the beginning of each lead, so these characters should be observed with particular care. The only exception to this occurs when the 'b' lead reads 'Combination of characters not as above'. In such cases, the specimen must agree with *all* of the characters given in the 'a' lead; if it deviates in one or more characters it must be treated as falling into the 'combination of characters not as above' category.

Experience has shown that users make errors more often in the key to groups than anywhere else; and, of course, an error here means that a correct identification is virtually impossible. The following paragraphs consist of a commentary on the key to the groups, with precise indications of how it should be used. the principles covered by this commentary, if not the details, are also relevant to the rest of the key.

Couplet 1 discriminates between the Dicotyledons and the Monocotyledones, the two large groups into which the Flowering Plants (Angiospermae) naturally divide. There is no single character that completely and certainly distinguishes these two groups; instead, a combination of a rather large number of characters has to be used, and couplet 1 includes the most readily observed of these. 1a starts with the phrase 'Cotyledons usually 2, lateral', as opposed to 1b, 'Cotyledon 1, terminal'. Specimens with 2 cotyledons (seedling leaves) clearly belong to 1a, but those with only 1 cotyledon could fit either lead, although it must be borne in mind that among the Dicotyledons the occurrence of species with only a single cotyledon is extremely rare. Of course, the chances of the user being able to answer this question if she has only a mature specimen are very small, but the character is important enough to be worth mentioning. The second phrases of the leads of the couplet are: 'leaves usually net-veined, with or without stipules, alternate, opposite or whorled' as opposed to 'leaves usually with parallel veins, sometimes these connected by cross-veinlets; leaves without stipules, opposite only in some aquatic plants'. These are admittedly very imprecise alternatives, but they do help in distinguishing the two groups. Firstly, if the specimen has stipules (or the scars left

[46]

by them when they have fallen), or if it is a terrestrial plant with opposite or whorled leaves, then clearly it matches with 1a. If the leaves are net-veined, then there is a high probability that the plant matches 1a, just as, if the leaves have parallel veins, there is a high probability that it goes with 1b.

The third characteristic used in this couplet reads: 'flowers with parts in 2s, 4s or 5s, or parts numerous' as opposed to 'flowers with parts in 3s'. 'Parts' here, essentially means sepals and petals or perianth, and stamens. Again, this raises a matter of probabilities: if the flower has parts in 2s, 4s, or 5s (or multiples of these), or the parts are numerous (more than 10), then there is a high probability that the plant belongs with 1a; if the parts are in 3s (or multiples, then there is a high probability that it belongs with 1b. Finally, the phrase 'primary root system (taproot) usually persistent, branched' as opposed to 'mature root-system wholly adventitious' again poses a similar set of probabilities.

In making a decision about a particular plant, it is necessary to observe what information is available and then to make a judgement as to the balance of probabilities between 1a and 1b. Fortunately, with experience, the discrimination of these two large groups is actually easier than it appears, and little difficulty is generally found in deciding whether a particular plant belongs to one or the other.

Couplet 2 (arrived at from 1a) is much more straightforward. If the plant has petals which are united to each other into a cup or tube at the base, then the plant matches 2b. If, on the other hand, the petals are quite separate from each other at the base (and fall individually as the flower dies), or if the flower has no petals at all, then the plant matches 2a. Some problems may arise in deciding whether or not there are petals present: these problems are discussed under 'Examining the Plant' on p. 28.

Couplet 3 (arrived at from 2a) separates off what are essentially catkin-bearing trees and shrubs. Plants matching 3a are always woody and have unisexual flowers with at least the males in catkins: inflorescences of small flowers without petals which are borne in the axils of bracts which overlap and protect the flowers. The anthers (and the stigmas of the female flowers, if these are also in catkins) project between the bracts when they are ripe (or receptive). If the plant is woody and such inflorescences are present, then the plant matches 3a; if not, then 3b should be chosen.

Couplet 4 concerns only the ovary. If it is made up of 2 or more separate carpels, then the specimen matches 4a. If, on the other hand, the carpels are joined to each other (even if only by the styles) then the specimen matches 4b. If the flower contains only a single carpel (which will have a single style), then it also goes with 4b.

Couplet 5 deals essentially with the question of whether the perianth is single or double (i.e. consists of a single whorl or of 2 or more distinct whorls — calyx and corolla). This is occasionally difficult to decide; see pp. 29–30 for more details. Having studied the perianth of the plant under consideration carefully, then a decision can be made as to whether the specimen matches 5a or 5b. The note about aquatic plants mentioned under 5a must also be considered: such plants are definitely excluded from 5a.

Couplet 6 is relatively simple. If there are more than twice as many stamens as petals, then 6a is chosen. If the stamens are twice as many as the petals or fewer, then 6b applies.

Couplet 7 is also relatively simple, requiring a decision as to whether the ovary of the flower is superior or inferior. This question is dealt with in detail on pp. 32–34.

Couplet 8 requires answers about the placentation of the ovary: see pp. 22–27 for details.

Couplet 9 (which is arrived at from couplet 5) is the equivalent of couplet 7, but is worded slightly differently to allow for the identification of plants with unisexual flowers, which occur with some frequency in the groups deriving from couplet 9.

Couplet 10 covers the same ground as couplets 7 and 9.

Couplet 11 is one that can lead to problems. The symmetry of the flowers is discussed in details on p. 30.

Couplet 12 (arrived at from couplet 1) deals with the Monocotyledons, breaking them down in to two large groups on the basis of whether the ovary is superior or inferior. The qualifying phrases to each couplet should be noted if the plant to be identified is aquatic.

[48]

Proceeding through the keys in this manner, making careful observations of the plant in the light of what is understood from a detailed reading of the various couplets, should lead to accurate identifications.

Keys

Key to Groups

1a   Cotyledons usually 2, lateral; leaves usually net-veined, with or without stipules, alternate, opposite or whorled; flowers with parts in 2s, 4s or 5s, or parts numerous; primary root-system (taproot) usually persistent, branched (*Dicotyledons*)   2

  b   Cotyledon 1, terminal; leaves usually with parallel veins, sometimes these connected by cross-veinlets; leaves without stipules, opposite only in some aquatic plants; flowers with parts in 3s; mature root-system wholly adventitious (*Monocotyledons*)   12

2a   (1) Petals present, free from each other at their bases (rarely united above the base), usually falling as individual petals, or petals absent   3

  b   Petals present and all united at the base into a longer or shorter tube, usually falling as a complete corolla   10

3a   (2) Flowers unisexual and without petals, at least the males borne in catkins which are usually deciduous; plants always woody

                                             Group A (p. 51)

  b   Flowers with or without petals, unisexual or bisexual, never in catkins; plants woody or not   4

4a   (3) Ovary consisting of 2 or more carpels which are completely free from each other, their styles also completely free from each other   **Group B** (p. 52)

  b   Ovary consisting of a single carpel or of 2 or more carpels which are united to each other wholly or in the greater part, rarely the bodies of the carpels free but the styles completely united   5

5a   (4) Perianth of 2 or more whorls, more or less clearly differentiated into calyx and corolla (calyx rarely small and obscure; excluding aquatic plants with minute, quickly deciduous petals and branch parasites with opposite, leathery leaves)   6 b Perianth of a single whorl (which may be corolla-like) or perianth completely absent, more rarely the perianth of 2 or more whorls but the segments not or scarcely differing from whorl to whorl   9

6a (5) Stamens more than twice as many as the petals
   Group C (p. 55)

b Stamens twice as many as the petals or fewer   7

7a (6) Ovary partly or fully inferior   Group D (p. 59)

b Ovary completely superior   8

8a (7) Placentation axile, apical, basal, marginal or free-central
   Group E (p. 61)

b Placentation parietal   Group F (p. 67)

9a (5) Stamens borne on the perianth or ovary inferior (perianth of female flowers sometimes very small)   Group G (p. 68)

b Stamens free from the perianth, ovary superior or naked (i.e. not surrounded by a perianth)   Group H (p. 71)

10a (2) Ovary partly or fully inferior   Group I (p. 74)

b Ovary completely superior   11

11a (10) Corolla actinomorphic   Group J (p. 76)

b Corolla zygomorphic   Group K (p. 81)

12a (1) Ovary superior or flowers completely without perianth (including all aquatics with totally submerged flowers)
   Group L (p. 84)

b Ovary partly or fully inferior (if plants aquatic then flowers borne well above water-level)   Group M (p. 88)

GROUP A

Flowers unisexual, at least the males borne in catkins which are erect or pendulous.

1a Stems jointed; leaves reduced to whorls of scales   1 **Casuarinaceae**

b Stems not jointed; leaves not as above   2

2a (1) Leaves pinnate   3

b Leaves simple and entire, toothed or lobed (sometimes deeply so   4

3a (2) Leaves without stipules; fruit a nut   3 **Juglandaceae**

b Leaves with stipules; fruit a legume   84 **Leguminosae**

4a (2) Male flowers with a perianth of 2 segments and 4–5 fertile stamens plus 4–5 staminodes; female flowers without a perianth; shrubs; fruit a syncarp of berries   **71 Bataceae**

  b Combination of characters not as above   5

5a (4) Leaves opposite, evergreen, entire; fruit berry-like

                                                    162 Garryaceae

  b Leaves alternate, deciduous or evergreen; fruit not berry- like   6

6a (5) Ovules many, parietal; seeds many, cottony-hairy; male catkin erect with the stamens projecting between the bracts, or hanging and with fringed bracts   **5 Salicaceae**

  b Ovules solitary or few, not parietal; seeds few, not cottony-hairy; male catkins not as above   7

7a (6) Leaves dotted with aromatic glands   **2 Myricaceae**

  b Leaves not dotted with aromatic glands   8

8a (7) Styles 3, each often branched; fruit splitting into 3 mericarps; seeds with appendages   **94 Euphorbiaceae**

  b Styles 1–6, not branched; fruit and seeds not as above   9

9a (8) Plants with milky sap   **10 Moraceae**

  b Plants with clear sap   10

10a (9) Male catkins compound, i.e. each bract with 2–3 flowers attached to it; styles 2   **6 Betulaceae**

  b Male catkins simple, i.e. each bract with a single flower attached to it; styles 1 or 3–6   11

11a (10) Ovary inferior; fruit a nut surrounded or enclosed by a scaly cupule; stipules deciduous; styles 3–6   **7 Fagaceae**

  b Ovary superior; fruit a leathery drupe, cupule absent; stipules absent; style 1   **4 Leitneriaceae**

GROUP B

Flowers unisexual or bisexual, none in catkins, ovary consisting of 2 or more carpels which are completely free from each other, each carpel with a single style which is free from all the other styles.

1a Trees with bark peeling off in plates; leaves palmately lobed; flowers unisexual in hanging, spherical heads   **72 Platanaceae**

  b Combination of characters not as above   2

2a (1) Perianth-segments and stamens borne independently below the
ovary, or perianth absent (i.e. ovary superior or naked)   3

b Perianth-segments and stamens borne on a rim or cup which is
itself borne below the ovary   21

3a (2) Aquatic plants with peltate leaves and flowers with 3 sepals
45 **Nymphaeaceae**

b Terrestrial plants, or if aquatic, then without peltate leaves and
flowers with more than 3 sepals   4

4a (3) Herbs, succulent shrubs or shrubs with yellow wood, or
climbers with bisexual flowers and opposite leaves   5

b Trees or shrubs which are neither succulent nor with yellow
wood, if climbers then with unisexual flowers and alternate leaves
10

5a (4) Perianth absent   47 **Saururaceae**

b Perianth present   6

6a (5) Leaves succulent; stamens in 1 or 2 whorls   74 **Crassulaceae**

b Leaves not succulent; stamens spirally arranged, not obviously in
whorls   7

7a (6) Petals fringed; fruits formed from each carpel borne on a
common stalk (gynophore)   69 **Resedaceae**

b Petals (when present) not fringed, but sometimes modified for
nectar-secretion; fruits formed from each carpel not borne on a
common stalk   8

8a (7) Leaves opposite or whorled; flowers small, stalkless, in axillary
clusters; ovule 1, placentation basal   17 **Phytolaccaceae**

b Combination of characters not as above   9

9a (8) Sepals differing among themselves, green; stamens ripening
from the inside of the flower outwards, borne on a
nectar-secreting disc   53 **Paeoniaceae**

b Sepals all similar, green or petal-like; stamens ripening from the
outside of the flower inwards; nectar-secreting disc absent
41 **Ranunculaceae**

10a (4) Leaves simple   11

b Leaves compound   20

11a (10) Sepals and petals 5 each   12

b Sepals and petals not 5 each   13

12a (11) Leaves opposite; stamens 5–10   104 **Coriariaceae**

    b Leaves alternate; stamens more than 10   **52 Dilleniaceae**
13a (11) Unisexual climbers   14
    b Erect trees or shrubs, usually flowers bisexual   15
14a (13) Carpels many; seeds not U-shaped   **32 Schisandraceae**
    b Carpels 3 or 6; seeds usually U-shaped   **44 Menispermaceae**
15a (13) Stamens each with a truncate connective which overtops the
    anther; fruit usually fleshy, formed from the closely contiguous
    products of several carpels (i.e. a syncarp); endosperm convoluted
         29 Annonaceae
    b Connectives of stamens not as above; fruit not as above;
    endosperm not convoluted   16
16a (15) Carpels spirally arranged on an elongate receptacle stipules
    large, united, early-deciduous leaving a ring-like scar
         27 Magnoliaceae
    b Carpels in 1 whorl; stipules absent, minute, or united to the
    leaf-stalk, not leaving a ring-like scar when fallen   17
17a (16) Petals present   18
    b Petals absent   19
18a (17) Sepals free, overlapping, more than 6; ovule solitary in each
    carpel   **33 Illiciaceae**
    b Sepals 2–6, united, or if free, then edge-to-edge in bud; ovules
    more than 1 in each carpel   **28 Winteraceae**
19a (17) Leaves in whorls; flowers bisexual; sepals minute or absent
         39 Eupteleaceae
    b Leaves opposite or alternate; flowers unisexual; sepals 4
         40 Cercidiphyllaceae
20a (10) Unisexual climbers, or erect shrubs with blue fruits; perianth
    parts in 3s   **43 Lardizabalaceae**
    b Erect shrubs, fruits not blue; perianth parts not in 3s
         53 Paeoniaceae
21a (2) Leaves modified into insect-trapping pitchers
         75 Cephalotaceae
    b Leaves not modified into pitchers   22
22a (21) Flowers unisexual; leaves evergreen   **34 Monimiaceae**
    b Flowers bisexual; leaves usually deciduous   23
23a (22) Inner stamens sterile; perianth of many segments; leaves
    usually opposite   **35 Calycanthaceae**

b Stamens all fertile; perianth of 4—9 segments; leaves usually
alternate   24
24a (23) Leaves without stipules, entire; flowers solitary and terminal;
seed with a divided aril   **54   Crossosomataceae**
b Leaves usually with stipules and toothed (if entire then flowers
clustered); seed without an aril   **82   Rosaceae**

GROUP C

*Flowers unisexual or bisexual, not borne in catkins; sepals and petals
both present; stamens more than twice as many as petals; ovary of a
single carpel or 2 or more united carpels, or, if the bodies of the
carpels are apparently free, then united by a common style.*

1a Herbaceous climber; leaves palmately divided into stalked leaflets;
petals 2, stamens 8   **90   Tropaeolaceae**
b Combination of characters not as above   2
2a (1) Perianth and stamens hypogynous, borne independently below
the superior ovary   3
b Perianth and stamens either perigynous, borne on the edge of a
rim or cup which itself is borne below the superior ovary, or
epigynous, borne on the top or the sides of the partly or fully
inferior ovary   35
3a (2) Placentation axile or free-central   4
b Placentation marginal or parietal   23
4a (3) Placentation free-central; sepals 2   **20   Portulacaceae**
b Placentation axile; sepals usually more than 2   5
5a (4) Leaves all basal, tubular, forming insect-trapping pitchers; style
peltately dilated   **63   Sarraceniaceae**
b Leaves not as above; style not peltately dilated   6
6a (5) Leaves alternate   7
b Leaves opposite or rarely whorled   20
7a (6) Anthers opening by terminal pores   8
b Anthers opening by slits   10
8a (7) Shrubs with simple leaves without stipules, often covered with
stellate hairs; stamens inflexed in bud; fruit a berry

56   Actinidiaceae

[55]

| | b | Combination of characters not as above | 9 |

9a (8) Ovary deeply lobed, borne on an enlarged receptacle or gynophore; petals not fringed   **57 Ochnaceae**

  b Ovary not lobed, not borne as above; petals often fringed   **123 Elaeocarpaceae**

10a (7) Perianth-segments of inner whorl tubular or bifid, nectar-secreting; fruit a group of partly to fully coalescent follicles   **41 Ranunculaceae**

  b Combination of characters not as above   11

11a (10) Leaves with translucent aromatic glands   **96 Rutaceae**

  b Leaves without such glands   12

12a (11) Sap milky; flowers unisexual   **94 Euphorbiaceae**

  b Sap not milky; flowers bisexual   13

13a (12) Succulent herb with spines; bark hard and resinous; stamens 15 in groups of 3 of which the central is the largest   **89 Geraniaceae**

  b Combination of characters not as above   14

14a (13) Large tropical trees; sepals 5, all or 2–3 of them enlarged and wing-like in fruit   **58 Dipterocarpaceae**

  b Combination of characters not as above   15

15a (14) Stipules absent; leaves evergreen   **59 Theaceae**

  b Stipules present; leaves usually deciduous   16

16a (15) Filaments free; anthers 2-celled   17

  b Filaments united into a tube, at least around the ovary, often also around the style; anthers often 1-celled   18

17a (16) Nectar-secreting disc absent; stamens more than 15; leaves simple   **124 Tiliaceae**

  b Nectar-secreting disc present, conspicuous; stamens 15; leaves dissected   **91 Zygophyllaceae**

18a (16) Styles divided above, several; stipules often persistent; carpels 5 or more   **125 Malvaceae**

  b Style 1, stigma capitate or lobed, 1–several; stipules usually deciduous; carpels 2–5   19

19a (18) Stamens in 2 whorls, those of the outer whorl usually sterile   **127 Sterculiaceae**

  b Stamens in several whorls, all fertile   **126 Bombacaceae**

20a (6) Sepals united, falling as a unit; fruit separating into

	boat-shaped units	**55**	**Eucryphiaceae**
b	Sepals and fruit not as above		21

21a (20) Small trees; stamens with brightly coloured filaments which are at least twice as long as the petals, the anthers forming a circle **60 Caryocaraceae**

  b Combination of characters not as above   22

22a (21) Leaves simple, without stipules, often with translucent glands; stamens often united in bundles **62 Guttiferae**

  b Leaves pinnate, without translucent glands; stamens not united in bundles **91 Zygophyllaceae**

23a (3) Aquatic plants with cordate leaves; style and stigmas forming a disc on top of the ovary **45 Nymphaeaceae**

  b Combination of characters not as above   24

24a (23) Carpel 1 with marginal placentation   25

  b Carpels 2 or more, placentation parietal   26

25a (24) Leaves bipinnate or modified into phyllodes, with stipules

84 Leguminosae

  b Leaves various but not as above, without stipules

41 Ranunculaceae

26a (24) Leaves modified into active insect-traps, the 2 halves of the blade fringed and closing rapidly when stimulated

65 Droseraceae

  b Combination of characters not as above   27

27a (26) Leaves opposite   28

  b Leaves alternate   30

28a (27) Styles numerous; floral parts in 3s **66 Papaveraceae**

  b Styles 1–5; floral parts in 4s or 5s   29

29a (28) Style 1; stamens not united in bundles; leaves without translucent glands **135 Cistaceae**

  b Styles 3–5, free or variously united below; stamens united in bundles (sometimes apparently free); leaves with translucent or blackish glands **62 Guttiferae**

30a (28) Small trees with aromatic bark; filaments of the stamens united **31 Canellaceae**

  b Herbs, shrubs or trees, bark not aromatic; filaments free   31

31a (29) Trees; leaves with stipules; anthers opening by short, pore-like slits **136 Bixaceae**

b	Herbs or shrubs; leaves usually without stipules; anthers opening by longitudinal slits	32
32a	(31) Sepals 2 or rarely 3, quickly deciduous	66 **Papaveraceae**
b	Sepals 4–8, persistent in flower	33
33a	(32) Leaves scale-like; styles 5	137 **Tamaricaceae**
b	Leaves not as above; styles 1, 2 or 3	34
34a	(33) Ovary closed at the apex, borne on a stalk (gynophore); none of the petals fringed	67 **Capparaceae**
b	Ovary open at apex, not borne on a stalk; at least some of the petals fringed	69 **Resedaceae**
35a	(2) Flowers unisexual; leaf bases oblique	143 **Begoniaceae**
b	Flowers bisexual; leaf bases not oblique	36
36a	(35) Placentation free-central; ovary partly inferior	20 **Portulacaceae**
b	Placentation not free-central; ovary either completely superior or completely inferior	37
37a	(36) Aquatic plants with cordate leaves	45 **Nymphaeaceae**
b	Terrestrial plants; leaves various	38
38a	(37) Carpels 1 or 3, eccentrically placed at the top of, the bottom of, or within the tubular perigynous zone	83 **Chrysobalanaceae**
b	Carpels and perigynous zone not as above	39
39a	(38) Stamens united into bundles on the same radii as the petals; staminodes often present; plants usually rough with stinging hairs	141 **Loasaceae**
b	Combination of characters not as above	40
40a	(39) Sepals 2, united, falling as a unit as the flower opens; plants herbaceous	66 **Papaveraceae**
b	Sepals 4–5, usually free, not falling as a unit; mostly trees or shrubs	41
41a	(40) Stamens united into several rings or sheets	149 **Lecythidaceae**
b	Stamens not as above	42
42a	(41) Carpels 8–12, superposed	148 **Punicaceae**
b	Carpels fewer, side-by-side	43
43a	(42) Leaves with stipules	44
b	Leaves without stipules	45
44a	(43) Leaves opposite or in whorls; plants woody	77 **Cunoniaceae**

b Leaves alternate; plants woody or herbaceous   82 **Rosaceae**
45a (43) Leaves with translucent aromatic glands; style 1

147 **Myrtaceae**

  b Leaves without such glands; styles usually more than 1

76 **Saxifragaceae**

GROUP D

*Flowers unisexual or bisexual, not borne in catkins; sepals and petals
both present; stamens twice as many as petals or fewer; ovary of a
single carpel or 2 or more united carpels, inferior.*

1a Petals and stamens numerous; plant succulent   2
  b Petals and stamens each 10 or fewer; plants usually not succulent

3

2a (1) Stems succulent, often very spiny; leaves usually absent, very
     reduced or falling early   26 **Cactaceae**
  b Stems and leaves succulent, spines usually absent   19 **Aizoaceae**
3a (1) Anthers opening by terminal pores   4
  b Anthers opening by longitudinal slits or by valves   5
4a (3) Filaments each with a knee-like joint below the anther; leaves
     usually with 3 conspicuous main veins from near the base

150 **Melastomataceae**

  b Filaments straight; leaves each with a single main vein

151 **Rhizophoraceae**

5a (3) Placentation parietal, placentas sometimes intrusive   6
  b Placentation axile, basal, apical or free-central   7
6a (5) Climbing herbs with tendrils; flowers unisexual

144 **Cucurbitaceae**

  b Erect herbs or shrubs, if climbing, then without tendrils; flowers
     usually bisexual   76 **Saxifragaceae**
7a (5) Stamens as many as and on the same radii as the petals; trees
     or shrubs with simple leaves   120 **Rhamnaceae**
  b Stamens more numerous than petals, or, if as numerous, then not
     on the same radii as them; plants herbaceous or woody, leaves
     simple or compound   8
8a (7) Flowers borne in umbels, these sometimes condensed into

	heads or in superposed whorls; leaves usually compound		9
b	Flowers not in umbels; leaves usually simple		10

9a (8) Fruit splitting into 2 mericarps; flowers usually bisexual; petals imbricate in bud; usually herbs without stellate hairs

164 **Umbelliferae**

  b Fruit a berry; flowers often unisexual; petals valvate in bud; plants mostly woody, often with stellate hairs   163 **Araliaceae**

10a (8) Style 1   11

  b Styles more than 1, often 2 and divergent   18

11a (10) Floating aquatic herb; leaf-stalks inflated   146 **Trapaceae**

  b Terrestrial herbs, trees or shrubs; leaf-stalks not inflated   12

12a (11) Small, low shrubs with scale-like, overlapping leaves; flowers in heads   81 **Bruniaceae**

  b Trees, shrubs or herbs with expanded leaves; flowers not usually in heads   13

13a (12) Ovary 1-celled with 2–5 ovules; fruit leathery or drupe-like, 1-seeded   152 **Combretaceae**

  b Ovary usually with 2–5 cells, ovules various; fruit not as above   14

14a (13) Ovule 1 in each cell of the ovary   15

  b Ovules 2–numerous in each cell of the ovary   17

15a (14) Petals imbricate in bud; flowers often unisexual

159 **Nyssaceae**

  b Petals valvate in bud; flowers usually bisexual   16

16a (15) Stamens with swollen, hairy filaments; petals recurved

158 **Alangiaceae**

  b Stamens without swollen, hairy filaments; petals not recurved

161 **Cornaceae**

17a (14) Sap milky; petals 5; ovary 3-celled   215 **Campanulaceae**

  b Sap watery; petals 2 or 4; ovary usually 4-celled   153 **Onagraceae**

18a (10) Stipules absent   76 **Saxifragaceae**

  b Stipules present, though sometimes early deciduous   19

19a (18) Ovary with half or more superior and half or less inferior; leaves opposite   77 **Cunoniaceae**

  b Ovary mostly or completely inferior; leaves alternate   20

20a (19) Anthers opening by valves; cells of ovary as many as styles; stellate hairs frequent   73 **Hamamelidaceae**

b Anthers opening by slits; cells of ovary ultimately twice as many
   as styles; stellate hairs absent   **82   Rosaceae**

GROUP E

*Flowers unisexual or bisexual, not borne in catkins; sepals and petals
both present; stamens twice as many as petals or fewer; ovary of a
single carpel or 2 or more united carpels, superior; placentation axile,
apical, basal, marginal or free-central.*

1a Perianth zygomorphic   2
 b Perianth actinomorphic (the stamens sometimes not actinomorphic
    due to deflexion)   15
2a (1) Anthers cohering above the ovary like a cap

                                           111   Balsaminaceae
 b Anthers free, not as above   3
3a (2) Anthers opening by terminal pores   4
 b Anthers opening by longitudinal slits or by valves   5
4a (3) Stamens 8, filaments united for at least half their length; fruit
    without barbed bristles   **103   Polygalaceae**
 b Stamens 3 or 4, filaments free; fruit covered with barbed bristles

                                            85   Krameriaceae
5a (3) Plants herbaceous   6
 b Plants woody (shrubs, trees or climbers)   10
6a (5) Leaves with stipules   7
 b Leaves without stipules   8
7a (6) Carpel 1; fruit a legume, sometimes 1-seeded   **84 Leguminosae**
 b Carpels 5; fruit a capsule or berry, or splitting into mericarps

                                            89   Geraniaceae
8a (6) Sepals, petals and stamens borne independently below the
    ovary (rarely the petals and stamens somewhat united at the
    base); leaves peltate   **90   Tropaeolaceae**
 b Sepals, petals and stamens borne on a rim, cup or tube which
    itself is borne below the ovary; leaves not peltate   9
9a (8) Leaves opposite   **145   Lythraceae**
 b Leaves alternate or all basal   **76   Saxifragaceae**

10a (5) Stamens as many as or fewer than petals, borne on the same radii as them   109 Sabiaceae

b Stamens more numerous than the petals, or if as many or fewer, not on the same radii as them   11

11a (10) Carpel 1   12

b Carpels 2 or more   13

12a (11) Style arising from near the base of the carpel
                                          83 Chrysobalanaceae

b Style arising from the apex of the carpel   84 Leguminosae

13a (11) Leaves opposite, palmate; sepals united at the base
                                          108 Hippocastanaceae

b Leaves alternate, usually pinnate; sepals free   14

14a (13) Stipules large, borne between the bases of the leaf-stalks
                                          110 Melianthaceae

b Stipules absent, or if present, not borne as above
                                          107 Sapindaceae

15a (1) Anthers opening by terminal pores   16

b Anthers opening by longitudinal slits or by valves   23

16a (15) Leaves with 3 more or less parallel veins from near the base; each filament with a knee-like joint below the anther
                                          150 Melastomataceae

b Leaves with 1 main vein; filaments not as above   17

17a (16) Low shrubs; leaves and stems covered with conspicuous stalked glandular hairs on which insects are often caught   18

b Shrubs or rarely low shrubs, not glandular-hairy as above   19

18a (17) Carpels 2   79 Byblidaceae

b Carpels 3   80 Roridulaceae

19a (17) Low shrubs with unisexual flowers; stamens 4, petals 4 each usually 2–3-lobed, rarely a few unlobed   123 Elaeocarpaceae

b Combination of characters not as above   20

20a (19) Carpels 2; leaves opposite   102 Tremandraceae

b Carpels 3 or more; leaves alternate   21

21a (20) Carpels 3; style divided above into 3 stigmas; nectar-secreting disc absent   166 Clethraceae

b Carpels usually 4 or 5, very rarely 3; style undivided or divided into 4 or 5 branches; nectar-secreting disc present around the base of the ovary   22

22a (21) Petals about as long as broad, not clawed; evergreen herbs or low shrubs; style divided above into 4–5 stigmas, rarely unlobed; seeds each with a wing-like projection at each end

167   **Pyrolaceae**

b Petals usually longer than broad, somewhat clawed; evergreen or deciduous shrubs; style undivided, stigmas as many as carpels, borne within a cup-like sheath; seeds various, rarely as above

168   **Ericaceae**

23a (15) Placentation free-central (ovary rarely with septa below) or basal   24

b Placentation axile or apical   29

24a (23) Stamens as many as petals and on the same radii as them   25

b Stamens more or fewer than petals, if as many then not on the same radii as them   28

25a (24) Small evergreen shrubs; leaves with translucent dots: flowers unisexual, sepals and petals 4; fruit a 1-seeded drupe

172   **Myrsinaceae**

b Combination of characters not as above   26

26a (25) Anthers opening by valves; stigma 1   42   **Berberidaceae**

b Anthers opening by longitudinal slits; stigmas more than 1   27

27a (26) Sepals 5; ovule 1, basal on a long, curved stalk

174   **Plumbaginaceae**

b Sepals 2 or rarely 3; ovules usually numerous, rarely 1 and then not on a long curved stalk   20   **Portulacaceae**

28a (24) Ovary lobed, consisting of several rounded humps, the style arising from a depression between them; leaves pinnatisect

87   **Limnanthaceae**

b Ovary not lobed, style terminal; leaves simple, entire

22   **Caryophyllaceae**

29a (23) Petals and stamens both numerous; plants with succulent leaves and stems   19   **Aizoaceae**

b Combination of characters not as above   30

30a (29) Small hairless annual herb growing in water or on wet mud; seeds pitted   139   **Elatinaceae**

b Combination of characters not as above   31

31a (30) Sepals, petals and stamens borne on a rim, cup or tube which itself is inserted below the ovary   32

b Sepals, petals and stamens inserted individually below the ovary (rarely the stamens united to the bases of the petals) 38

32a (31) Stamens as many as the petals and borne on the same radii as them **120 Rhamnaceae**

b Stamens more or fewer than the petals, or if of the same number, then not on the same radii as them 33

33a (32) Style 1 34

b Styles more than 1, often 2 and divergent 35

34a (33) Perigynous zone not prominently ribbed; seeds with arils; mostly trees, shrubs or climbers **115 Celastraceae**

b Perigynous zone prominently ribbed; seeds without arils; mostly herbs **145 Lythraceae**

35a (33) Fruit an inflated, membranous capsule; leaves mostly opposite, compound **116 Staphyleaceae**

b Combination of characters not as above 36

36a Leaves opposite, leathery **77 Cunoniaceae**

b Leaves usually alternate or all basal, not leathery if opposite 37

37a (36) Trees or shrubs; hairs often stellate; anthers usually opening by valves; fruit a few-seeded, woody capsule **73 Hamamelidaceae**

b Herbs or shrubs; hairs simple or absent; anthers opening by longitudinal slits; fruit a capsule, not woody **76 Saxifragaceae**

38a (31) Leaves with translucent, aromatic glands **96 Rutaceae**

b Leaves without such glands 39

39a (38) Sap often milky; flowers unisexual; styles 3, often further divided **94 Euphorbiaceae**

b Combination of characters not as above 40

40a (39) Flower with a well-developed disc, usually nectar- secreting, below and around the ovary 41

b Flower without a disc, nectar secreted in other ways 51

41a (40) Stamens as many as and on the same radii as the petals 42

b Stamens more or fewer than the petals, or if of the same number not on the same radii as them 43

42a (41) Climbers with tendrils; stamens free **121 Vitaceae**

b Erect shrubs without tendrils; stamens with their filaments united, at least at the base **122 Leeaceae**

43a (41) Resinous trees or shrubs 44

b Herbs, shrubs or trees, not resinous, sometimes aromatic 45

44a (43) Ovules 2 in each cell of the ovary; fruit a drupe or capsule

                                                     99   Burseraceae

  b Ovule 1 in each cell of the ovary; fruit a drupe   105 Anacardiaceae

45a (43) Plant herbaceous   46

  b Plant woody (a tree, shrub or climber)   47

46a (45) Petals long-clawed, united above the base; leaves not fleshy

                                    117   Stackhousiaceae

  b Petals entirely free, not long-clawed; leaves fleshy

                                  91   Zygophyllaceae

47a (45) Flowers, or at least some of them, functionally unisexual (i.e.
    anthers not producing pollen or ovary containing no ovules)   48

  b Flowers functionally bisexual   49

48a (47) Leaves alternate; ovary with 2–5 carpels, not flattened

                                  98   Simaroubaceae

  b Leaves opposite; ovary with 2 (rarely 3) carpels, usually flattened

                                106   Aceraceae

49a (47) Leaves entire or toothed; stamens 4–5, emerging from the
    disc; seeds with arils   115   Celastraceae

  b Combination of characters not as above   50

50a (49) Leaves without stipules, not fleshy; filaments united into a
    tube   100   Meliaceae

  b Leaves with stipules, fleshy; filaments free   91   Zygophyllaceae

51a (40) Plant herbaceous   52

  b Plant woody (tree, shrub or climber)   54

52a (51) Leaves always simple; ovary 6–10-celled by the development
    of 3–5 secondary septa during maturation of the flower

                                  92   Linaceae

  b Leaves lobed or compound; secondary septa absent   53

53a (52) Leaves with stipules   89   Geraniaceae

  b Leaves without stipules   88   Oxalidaceae

54a (51) Filaments of stamens united below   55

  b Filaments of stamens entirely free from each other   59

55a (54) Plants succulent, spiny; stamens 8 with woolly filaments;
    plants unisexual   25   Didieriaceae

  b Combination of characters not as above   56

56a (55) Stamens 2   179   Oleaceae

  b Stamens 3 or more   57

57a (56) Leaves without stipules   **13   Olacaceae**

  b Leaves with stipules (though these sometimes quickly deciduous)

                                                       58

58a (57) Stipules persistent, borne between the bases of the leaf-stalks; petals with appendages   **93   Erythroxylaceae**

  b Stipules quickly deciduous, not borne as above; petals without appendages   **127   Sterculiaceae**

59a (54) Stamens 8–10   60

  b Stamens 3–6   62

60a (59) Petals long-clawed, often fringed or toothed; stamens 10; usually some or all of the sepals with nectar-secreting appendages on the outside   **101   Malpighiaceae**

  b Petals neither clawed nor fringed not toothed; stamens 8–10; sepals without such appendages   61

61a (60) Ovule 1 per cell; sepals united at their bases   **13   Olacaceae**

  b Ovules many per cell; sepals free   **132   Stachyuraceae**

62a (59) Staminodes present in flowers which also contain fertile stamens   63

  b Staminodes absent from flowers with fertile stamens, present only in female flowers   64

63a (62) Carpels 2, ovary containing a single, apical ovule; stamens as many as and on the same radii as the petals, attached to their bases; stipules present   **114   Corynocarpaceae**

  b Carpels 2–4, each cell of the ovary containing 1–2 ovules; stamens not on the same radii as the petals; stipules absent

                                    112   Cyrillaceae

64a (62) Trees with opposite, pinnate leaves; twigs tipped with large, dark buds; fruit a samara   **179   Oleaceae**

  b Combination of characters not as above   65

65a (64) Sepals united at the base   66

  b Sepals entirely free from one another   67

66a (65) Carpels 1 or rarely 3 with 1 or 2 of them sterile, the fertile containing 2 apical ovules   **119   Icacinaceae**

  b Carpels 3–many, all fertile, each containing 1–2 ovules

                                113   Aquifoliaceae

67a (65) Ovule 1 per cell; petals 3–4   **97   Cneoraceae**

  b Ovules many per cell; petals 5   **78   Pittosporaceae**

GROUP F

Flowers bisexual or unisexual, not in catkins; petals present;
stamens twice as many as the petals or fewer; ovary superior,
of a single carpel or several united carpels; placentation
parietal.

1a	Sepals, petals and stamens perigynous, borne on a rim or cup which itself is inserted below the ovary	2
b	Sepals, petals and stamens hypogynous, inserted individually below the ovary	5
2a	(1) Trees; leaves bi- or tripinnate; flowers bilaterally symmetric; stamens 5, of differing lengths	70 **Moringaceae**
b	Combination of characters not as above	3
3a	(2) Annual, aquatic herb; stamens 6	68 **Cruciferae**
b	Combination of characters not as above	4
4a	(3) Flower-stalks slightly united to the leaf-stalks so that the flowers appear to be borne on the latter; petals contorted in bud; carpels 3	133 **Turneraceae**
b	Flower-stalks not united to the leaf-stalks; petals not contorted in bud; carpels 2 or 4	76 **Saxifragaceae**
5a	(1) Perianth zygomorphic	6
b	Perianth actinomorphic	9
6a	(5) Ovary open at apex; some or all of the petals fringed	69 **Resedaceae**
b	Ovary closed at the apex; no petals fringed	7
7a	(6) Petals and stamens 5; carpels 2 or 3	131 **Violaceae**
b	Petals and stamens 4 or 6; carpels 2	8
8a	(7) Ovary borne on a stalk (gynophore); stamens projecting beyond petals	67 **Capparaceae**
b	Ovary not borne on a stalk; stamens not projecting beyond petals	66 **Papaveraceae**
9a	(5) Petals and stamens numerous	19 **Aizoaceae**
b	Petals and fertile stamens each fewer than 7	10
10a	(9) Stamens alternating with much-divided staminodes	76 **Saxifragaceae**
b	Stamens not alternating with much-divided staminodes	11

11a (10) Leaves insect-trapping and -digesting by means of stalked, glandular hairs   **65 Droseraceae**
  b Leaves not as above   12
12a (11) Climbers with tendrils; ovary and stamens borne on a common stalk (androgynophore); corona present

    134 Passifloraceae
  b Combination of characters not as above   13
13a (12) Petals 4, the outer pair trifid; sepals 2   **66 Papaveraceae**
  b Petals not as above; sepals 4–5   14
14a (13) Stamens usually 6, 4 longer and 2 shorter, rarely reduced to 2; carpels 2; fruit usually with a secondary septum

    68 Cruciferae
  b Stamens 4–10, all more or less equal; carpels 2–5, fruit without a secondary septum   15
15a (14) Petals each with a scale-like appendage at the base of the blade; leaves opposite   **138 Frankeniaceae**
  b Petals without appendages; leaves alternate or all basal   16
16a (15) Stipules present   **131 Violaceae**
  b Stipules absent   17
17a (16) Leaves alternate, scale-like   **137 Tamaricaceae**
  b Leaves usually all basal, normally developed   **167 Pyrolaceae**

GROUP G

Flowers bisexual or unisexual, not in catkins; perianth consisting of 1 whorl only, petals absent; stamens borne on the perianth or ovary inferior; ovary of a single carpel or 2 or more united carpels.

1a Aquatic plants or rhubarb-like marsh plants with cordate leaves   2
  b Terrestrial plants, not as above   5
2a (1) Stamens 8, 4 or 2; leaves either deeply divided or cordate at the base   **154 Haloragaceae**
  b Stamens 6 or 1; leaves undivided, not cordate at the base   3
3a (2) Stamens 6; leaves all basal   **68 Cruciferae**
  b Stamen 1; leaves opposite or in whorls   4
4a (3) Leaves in whorls; fruits small, indehiscent, dry, 1-seeded, not lobed   **156 Hippuridaceae**

b Leaves opposite; fruit 4-lobed with up to 4 seeds

194 **Callitrichaceae**

5a (1) Trees or shrubs   6

 b Herbs, climbers or parasites   22

6a (5) Flowers small, the central one bisexual, the rest male, in heads
   subtended by 2 large, white bracts   160 **Davidiaceae**

 b Inflorescence and flowers not as above   7

7a (6) Plant covered by scales; fruit enclosed in the berry-like,
   persistent, fleshy calyx   129 **Elaeagnaceae**

 b Plant not covered by scales; fruit not as above   8

8a (7) Stamen 1, or 1 complete stamen flanked by 2 half-stamens;
   leaves opposite, evergreen   49 **Chloranthaceae**

 b Stamens not as above; leaves not usually as above   9

9a (8): Stamens as many as, and on radii alternating with, the
   perianth-segments   120 **Rhamnaceae**

 b Stamens not arranged as above   10

10a (9) Stamens 4, situated at the tops of the spoon-shaped, petal-like
   perianth-segments   12 **Proteaceae**

 b Combination of characters not as above   11

11a (10) Stipules present, sometimes falling early   12

 b Stipules absent   16

12a (11) Ovary of a single carpel   13

 b Ovary of 2–6 united carpels   14

13a Stamens numerous, borne on the reflexed inner surface of the
   funnel-shaped perianth   82 **Rosaceae**

 b Stamens 10, perianth not as above, the stamens not borne on it

84 **Leguminosae**

14a (12) Styles 3–6; fruit a nut surrounded by a scaly cupule

7 **Fagaceae**

 b Styles 2; fruit not as above   15

15a (14) Leaves alternate; stellate hairs usually present; fruit a woody
   capsule   73 **Hamamelidaceae**

 b Leaves opposite; stellate hairs absent; fruit a non-woody capsule

77 **Cunoniaceae**

16a (11) Ovary of a single carpel   17

 b Ovary of 2–3 united carpels   20

17a (16) Ovary superior   18

b	Ovary inferior		19

18a (17) Leaves with aromatic glands; stamens 8 or more, all
borne at more or less the same level in the perigynous
zone   36   **Lauraceae**

  b Leaves without aromatic glands; stamens 2 or 8–10, when borne
at different levels in the perigynous zone   128   **Thymeleaceae**

19a (17) Epigynous zone present above the ovary, bearing the perianth
on its rim and the stamens in its inner surface; free-living trees
                                    152   **Combretaceae**

  b Epigynous zone absent; semi-parasitic shrubs   14   **Santalaceae**

20a (16) Ovary superior   106   **Aceraceae**

  b Ovary inferior   21

21a (20) Placentation parietal   76   **Saxifragaceae**

  b Placentation axile   161   **Cornaceae**

22a (5) Plants parasitic   23

  b Plants not parasitic   26

23a (22) Branch-parasites with green, forked branches or flowers
stalkless on branches of host   24

  b Root-parasites lacking chlorophyll   25

24a (23) Flowers borne on green, forked branches   15   **Loranthaceae**

  b Flowers brown, minute, unstalked   51   **Rafflesiaceae**

25a (23) Flowers minute in fleshy spikes; stamen 1
                                157   **Cynomoriaceae**

  b Flowers conspicuous in short, bracteate spikes   51   **Rafflesiaceae**

26a Perianth absent; flowers in spikes   47   **Saururaceae**

  b Perianth present; flowers not usually in spikes   27

27a (26) Leaf-base oblique; ovary inferior, 3-celled   143   **Begoniaceae**

  b Leaf-base not oblique; ovary not as above   28

28a (27) Ovary superior   29

  b Ovary inferior   34

29a (28) Carpel 1, containing a single apical ovule; perianth tubular
                              128   **Thymeleaceae**

  b Combination of characters not as above   30

30a (29) Carpels 3 (rarely 2), ovule 1, basal; perianth persistent in
fruit; leaves usually alternate, entire   31

  b Combination of characters not as above   32

31a (30) Leaves without stipules; stamens 5   21   **Basellaceae** b Leaves usually with stipules united into a sheath (ochrea); stamens usually 6–9   16   **Polygonaceae**

32a (30) Leaves alternate, usually lobed or compound   82   **Rosaceae**

  b Leaves opposite, usually entire   33

33a (32) Ovule 1, fruit a nut; stipules translucent and papery or rarely absent   22   **Caryophyllaceae**

  b Ovules numerous; fruit a capsule; stipules absent 145   **Lythraceae**

34a (28) Leaves pinnate; ovary open at apex   142   **Datiscaceae**

  b Leaves not pinnate; ovary closed at apex   35

35a (34) Ovary 6-celled; perianth 3-lobed or tubular and bilaterally symmetric   50   **Aristolochiaceae**

  b Combination of characters not as above   36

36a (35) Ovules 1–5; seed 1   37

  b Ovules and seeds numerous   38

37a (36) Perianth-segments thickening in fruit; leaves alternate

                                                    23   **Chenopodiaceae**

  b Perianth-segments not as above; leaves opposite or alternate

                                                    14   **Santalaceae**

38a (36) Styles 2; placentation parietal   76   **Saxifragaceae**

  b Style 1; placentation axile   153   **Onagraceae**

GROUP H

Flowers unisexual or bisexual; perianth of a single whorl of segments, not discriminated into calyx and corolla; stamens free from perianth; ovary of a single carpel or of 2 or more united carpels.

1a Aquatic plants, either submerged or at least partially covered by flowing water   2

  b Terrestrial plants, not as above   3

2a (1) Leaves whorled, much-divided   46   **Ceratophyllaceae**

  b Leaves not as above, plants sometimes not well differentiated into stems and leaves   86   **Podostemaceae**

3a (1) Climbers or scramblers, most leaves ending in a tendril-like structure which itself terminates in an insectionous pitcher

                                                    64   **Nepenthaceae**

b	Combination of characters not as above	4
4a	(3) Stipules present, sometimes falling early	5
b	Stipules entirely absent	14
5a	(3) Ovary 1-celled, containing a single ovule	6
b	Ovary 1–several-celled, containing more than a single ovule	9
6a	(5) Styles 2–4, usually 3, free	16 Polygonaceae
b	Style 1, sometimes divided above into 2 stigmas	7
7a	(6) Ovule basal; herbs or shrubs, flowers never sunk in a fleshy receptacle	11 Urticaceae
b	Ovule apical; trees, shrubs or woody climbers, if herbs then flowers sunk in a fleshy receptacle	8
8a	(7) Sap watery; style 1; flowers often bisexual	8 Ulmaceae
b	Sap milky; styles usually 2, rarely 1; flowers usually unisexual	10 Moraceae
9a	(5) Placentation parietal or free-central	10
b	Placentation axile	11
10a	(9) Shrubs or trees; leaves alternate; placentation parietal	130 Flacourtiaceae
b	Herbs; leaves usually opposite; placentation free-central	22 Caryophyllaceae
11a	(9) Sap milky; styles usually 3, often divided; ovules 1–2 per cell	94 Euphorbiaceae
b	Combination of characters not as above (ovules occasionally 2 per cell)	12
12a	(11) Stellate hairs usually present; stamens 5 or 10, filaments united below	127 Sterculiaceae
b	Stellate hairs absent; stamens not as above	13
13a	(12) Style 1; trees or shrubs	123 Elaeocarpaceae
b	Styles 3–4; herbs	47 Saururaceae
14a	(4) Ovary 1-celled, containing a single ovule	15
b	Ovary 1–several-celled, containing more than a single ovule	24
15a	(14) Tree with milky sap; styles 2	9 Eucommiaceae
b	Combination of characters not as above	16
16a	(17) Stamens with filaments united, at least at the base	17
b	Stamens with the filaments completely free	18
17a	(16) Aromatic trees; stigma 1; seed with a conspicuous aril	30 Myristicaceae

b Herbs or shrubs, not aromatic; stigmas 2–3; seeds without arils

24 Amaranthaceae

18a (16) Styles 2–5, completely free   19

  b Style 1 or style absent, stigmas sessile on the ovary   20

19a (18) Stamens 9; perianth petal-like, of 6 segments united below; flowers in an umbel or head partly enclosed in an involucre of 4–8-lobed bracts   16 Polygonaceae

  b Stamens usually 5; perianth not as above, generally of 2–5, free, sepal-like segments; inflorescence not as above   23 Chenopodiaceae

20a (18) Stigmas sessile on the ovary, brush-like; flowers sunk in fleshy spikes   48 Piperaceae

  b Combination of characters not as above   21

21a (20) Leaves opposite or in whorls   18 Nyctaginaceae

  b Leaves alternate   22

22a (21) Herb; stamens 7–22; perianth of 2 free segments

155 Theligonaceae

  b Shrubs; stamens 4–many; perianth of 4–6 segments united at least at the base   23

23a (22) Leaves with translucent, aromatic glands; stamens with enlarged connectives; fruit a syncarp   29 Annonaceae

  b Leaves without translucent aromatic glands; stamens without enlarged connectives; fruit a berry   17 Phytolaccaceae

24a (14) Styles 2 or more, free for all or most of their length   25

  b Style 1, sometimes lobed above into 2 or more stigmas   29

25a (24) Ovary of 2–3 cells   26

  b Ovary of 4 or more cells   26

26a (25) Ovary usually 3-celled (rarely 2-celled); ovules 1–2 per cell, all fertile   118 Buxaceae

  b Ovary 2-celled, though the septum is incomplete; ovules 2 per cell, but only 1 of the 4 developing into a seed

95 Daphniphyllaceae

27a (25) Ovules 1 per cell; shrubs or large herbs   17 Phytolaccaceae

  b Ovules many per cell; trees   28

28a (27) Leaves alternate; carpels 4   37 Tetracentraceae

  b Leaves in whorls; carpels 6–10   38 Trochodendraceae

29a (24) Placentation parietal or free-central   30

  b Placentation axile, apical or basal   31

30a (29) Placentation parietal; perianth-segments 2, free

66 **Papaveraceae**

b Placentation free-central; perianth-segments 5, united below

173 **Primulaceae**

31a (29) Leaves modified into insect-trapping pitchers

63 **Sarraceniaceae**

b Leaves not modified into insect-trapping pitchers 32

32a (32) Heath-like shrublets with narrow leaves with their margins
revolute; ovule 1 per cell 169 **Empetraceae**

b Plants not as above; ovules 2 or more per cell 34

33a (33) Leaves opposite; stamens 2 179 **Oleaceae**

b Leaves alternate; stamens 3 or more 33

34a (32) Resinous trees or shrubs; leaves simple or compound; fruit
1-seeded, drupe-like; ovule 1 per cell, apical or basal

105 **Anacardiaceae**

b Non-resinous trees, shrubs or climbers; leaves usually compound;
fruit not as above; ovules 2 per cell, axile 107 **Sapindaceae**

GROUP I

*Calyx and corolla both present, corolla made up of petals that are
united into a tube, at least below; stamens twice the number of
petals or fewer; ovary inferior.*

1a Leaves needle-like or scale-like; small, heather-like shrublets

81 **Bruniaceae**

b Combination of characters not as above 2

2a (1) Inflorescence a head surrounded by an involucre of bracts;
ovule always solitary 3

b Inflorescence and ovules not as above 4

3a (2) Each flower with a cup-like involucel; stamens 4, free; ovule
apical 214 **Dipsacaceae**

b Involucel absent; stamens 5, their anthers united into a tube;
ovule basal 219 **Compositae**

4a (2) Stamens 2; stamens and style united into a touch-sensitive
column; leaves linear 218 **Stylidiaceae**

b Combination of characters not as above 5

5a	(4) Leaves alternate or all basal	6
b	Leaves opposite or whorled	15
6a	(5) Anthers opening by pores; fruit a berry or drupe	
		168 **Ericaceae**
b	Anthers opening by slits; fruit various	7
7a	(6) Evergreen trees or shrubs; corolla white, campanulate; ovary half-inferior; placentation free-central, ovules few	172 **Myrsinaceae**
b	Combination of characters not as above	8
8a	(7) Climbers with tendrils and unisexual flowers; stamens 1–5; placentation parietal; fruit berry-like	144 **Cucurbitaceae**
b	Combination of characters not as above	9
9a	(8) Stamens 10–many; plants woody	10
b	Stamens 4–5; plants woody or herbaceous	12
10a	(9) Leaves gland-dotted, smelling of eucalyptus; corolla completely united, unlobed, falling as a whole	147 **Myrtaceae**
b	Combination of characters not as above	11
11a	(10) Hairs stellate or scale-like; stamens in 1 series, anthers linear	177 **Styracaceae**
b	Hairs absent or not as above; stamens in several series; anthers broad	178 **Symplocaceae**
12a	(9) Stigmas surrounded by a sheath	216 **Goodeniaceae**
b	Stigmas not surrounded by a sheath	13
13a	(12) Stamens as many as and on the same radii as the petals	173 **Primulaceae**
b	Stamens not as above	14
14a	(13) Stamens 2 or 4, borne on the corolla; sap not milky	205 **Gesneriaceae**
b	Stamens 5 or more, free from the corolla; sap usually milky	215 **Campanulaceae**
15a	(5) Placentation parietal; stamens 2, or 4 and paired	205 **Gesneriaceae**
b	Placentation axile or apical; stamens 1 or more, if 4 then not paired	16
16a	(15) Stamens 1–3; ovary with 1 ovule	213 **Valerianaceae**
b	Stamens 4 or more; ovary usually with 2 or more ovules	17
17a	(16) Leaves divided into 3 leaflets; flowers few in a head	212 **Adoxaceae**

b Leaves simple or rarely pinnate; inflorescence various, usually not as above   18

18a (17) Stipules usually borne between the bases of the leaf-stalks and sometimes looking like leaves; ovary usually 2-celled, more rarely 5-celled; flowers usually actinomorphic; fruit capsular, fleshy or schizocarpic   **186 Rubiaceae**

b Stipules usually absent, when present not as above; ovary usually 3-celled (occasionally 2–5-celled), sometimes only 1 cell fertile; flowers often zygomorphic; fruit a berry or drupe

211 Caprifoliaceae

GROUP J

Flowers usually bisexual; calyx and corolla both present, the corolla made up of petals which are united, at least at the base, actinomorphic; ovary of 1 carpel or of several carpels.

1a Stamens 2   **179 Oleaceae**

b Stamens more than 2   2

2a (1) Carpels several, free; leaves succulent   **74 Crassulaceae**

b Carpels united, or, if the bodies of the carpels are free, then the styles united, rarely ovary of a single carpel; leaves usually not succulent   3

3a (2) Corolla papery, translucent, 4-lobed; stamens 4, projecting form the corolla; leaves with parallel veins, often all basal

210 Plantaginaceae

b Combination of characters not as above   4

4a (3) Central flowers of the inflorescence abortive, their bracts forming nectar-secreting pitchers; petals completely united, the corolla falling as a whole as the flower opens   **61 Marcgraviaceae**

b Combination of characters not as above   5

5a (4) Stamens more than twice as many as the petals   6

b Stamens up to twice as many as the petals   13

6a (5) Leaves evergreen, divided into 3 leaflets; filaments brightly coloured, at least twice as long as the petals   **60 Caryocaraceae**

b Leaves deciduous or evergreen, simple, entire or lobed; filaments not as above   7

7a (6) Leaves with stipules; filaments of stamens united into a tube
around the ovary and style   125   Malvaceae

 b Leaves without stipules; filaments free   8

8a (7) Anthers opening by pores   56   Actinidiaceae

 b Anthers opening by longitudinal slits   9

9a (8) Leaves with translucent, aromatic glands; calyx cup-like,
unlobed   96   Rutaceae

 b Leaves without such glands; calyx not as above   10

10a (9) Placentation parietal; leaves fleshy   188   Fouquieriaceae

 b Placentation axile; leaves not fleshy   11

11a (10) Sap milky; ovules 1 per cell   175   Sapotaceae

 b Sap not milky; ovules 2 or more per cell   12

12a (11) Ovules 2 per cell; flowers usually unisexual   176   Ebenaceae

 b Ovules many per cell; flowers bisexual   59   Theaceae

13a (5) Stamens as many as petals and on the same radii as them   14

 b Stamens more or fewer than petals, if as many then not on the
same radii as them   21

14a (13) Tropical trees with milky sap and evergreen leaves

           175   Sapotaceae

 b Tropical or temperate trees, shrubs, herbs or climbers, with watery
sap and usually deciduous leaves   15

15a (14) Placentation axile   16

 b Placentation basal or free-central   17

16a (15) Climbers with tendrils; stamens free   121   Vitaceae

 b Upright shrubs without tendrils; stamens with their filaments
united below   122   Leeaceae

17a (15) Trees or shrubs; fruit a berry or drupe   18

 b Herbs (occasionally woody at the extreme base); fruit a capsule or
indehiscent   19

18a (17) Leaves with translucent glands; anthers opening towards the
centre of the flower; staminodes absent   172   Myrsinaceae

 b Leaves without such glands; anthers opening towards the outside
of the flower; staminodes 5   171   Theophrastaceae

19a (17) Sepals 2, free   20   Portulacaceae

 b Sepals 4 or more, united   20

20a (19) Corolla persistent and papery in fruit; ovule 1 on a long stalk
arising from the base of the ovary   174   Plumbaginaceae

  b Corolla not persistent and papery in fruit; ovules many, on a free-central placenta   173 **Primulaceae**

21a (13) Flower compressed with 2 planes of symmetry; stamens united in 2 bundles   66 **Papaveraceae**

  b Combination of characters not as above   22

22a (21) Leaves bipinnate or replaced by phyllodes; carpel 1; fruit a legume   84 **Leguminosae**

  b Combination of characters not as above   23

23a (22) Anthers opening by pores   24

  b Anthers opening by longitudinal slits, or pollen in coherent masses (pollinia)   25

24a (23) Stamens free from corolla-tube, often twice as many as the corolla-lobes   168 **Ericaceae**

  b Stamens attached to the corolla-tube, as many as the corolla-lobes   197 **Solanaceae**

25a (23) Leaves alternate or all basal; carpels never 2 and free or almost so but with a single terminal style   26

  b Leaves opposite or whorled, alternate only when carpels 2, free or almost so and style 1, terminal   42

26a (25) Plant woody, leaves usually evergreen, often spiny-margined; stigma not stalked, borne directly on the top of the ovary   113 **Aquifoliaceae**

  b Combination of characters not as above   27

27a (26) Procumbent herbs with milky sap and stamens free from the corolla-tube   215 **Campanulaceae**

  b Combination of characters not as above   28

28a (27) Ovary 5-celled   29

  b Ovary 2-, 3- or 4-celled   31

29a (28) Placentation parietal; soft-wooded tree   140 **Caricaceae**

  b Placentation axile; herbs   30

30a (29) Leaves fleshy; anthers 2-celled; fruit often deeply lobed, schizocarpic   196 **Nolanaceae**

  b Leaves leathery; anthers 1-celled; fruit a capsule or berry   170 **Epacridaceae**

31a (28) Ovary 3-celled   32

  b Ovary 1-, 2- or 4-celled   33

[78]

32a (31) Dwarf evergreen shrublets; 5 staminodes usually present; petals imbricate in bud   **165 Diapensiaceae**

  b Herbs or climbers with tendrils; staminodes absent; petals contorted in bud   **187 Polemoniaceae**

33a (31) Stamens with filaments united into a tube; flowers in heads; stigmas surrounded by a sheath   **217 Brunoniaceae**

  b Combination of characters not as above   34

34a (33) Flowers in spirally coiled cymes, or the calyx with appendages between the lobes; style terminal or arising from between the lobes of the ovary   35

  b Flowers not in spirally coiled cymes, calyx without appendages; style terminal   36

35a (34) Style terminal; fruit a capsule, usually many-seeded

    190 Hydrophyllaceae

  b Style arising from the depression between the 4 lobes of the ovary; fruit of up to 4 nutlets, or more rarely a 1–4-seeded drupe

    191 Boraginaceae

36a (34) Placentation parietal   37

  b Placentation axile   38

37a (36) Corolla-lobes valvate in bud; leaves simple and cordate or peltate, or of 3 leaflets, hairless; aquatic or marsh plants

    183 Menyanthaceae

  b Corolla-lobes imbricate in bud; leaves never as above; not aquatics or marsh plants   **205 Gesneriaceae**

38a (36) Ovules 1–2 in each cell of the ovary   39

  b Ovules 3–many in each cell of the ovary   41

39a (38) Arching shrubs with small purple flowers in clusters on the previous year's wood   **198 Buddlejaceae**

  b Combination of characters not as above   40

40a (39) Sepals free; corolla-lobes contorted and infolded in bud; twiners, herbs or dwarf shrubs   **189 Convolvulaceae**

  b Sepals united; corolla-lobes not as above in bud; trees or shrubs

    191 Boraginaceae

41a (39) Corolla-lobes folded, valvate or contorted in bud; septum of the ovary oblique, not in the horizontal plane   **197 Solanaceae**

  b Corolla lobes variously imbricate but not as above in bud; septum of ovary in the horizontal plane   **199 Scrophulariaceae**

42a	(25) Trailing, heather-like shrublet	**168 Ericaceae**
b	Plant not as above	43
43a	(42) Milky sap usually present; fruit usually of 2 almost free follicles united by a common style; seeds with silky appendages	44
b	Milky sap absent; fruit a capsule or fleshy, carpels united; seeds without silky appendages	45
44a	(43) Pollen granular; corona absent; corolla-lobes valvate in bud	**184 Apocynaceae**
b	Pollen usually in coherent masses (pollinia); corona usually present; corolla-lobes valvate or contorted in bud	**185 Asclepiadaceae**
45a	(43) Root-parasites without chlorophyll	**192 Lennoaceae**
b	Free-living plants with chlorophyll	46
46a	(45) Flowers in coiled cymes; usually herbs	**190 Hydrophyllaceae**
b	Flowers not in coiled cymes; herbs or shrubs	47
47a	(46) Placentation parietal; carpels 2	48
b	Placentation axile; carpels 2, 3 or 5	49
48a	(47) Leaves compound; epicalyx present	**190 Hydrophyllaceae**
b	Leaves simple; epicalyx absent	**182 Gentianaceae**
49a	(47) Stamens fewer than corolla-lobes	**193 Verbenaceae**
b	Stamens as many as corolla-lobes	50
50a	(49) Carpels 5; shrubs with leaves with spiny margins	**181 Desfontainiaceae**
b	Carpels 2 or 3; herbs or shrubs, leaves not as above	51
51a	(50) Leaves without stipules; carpels 3; corolla-lobes contorted in bud; herbs	**187 Polemoniaceae**
b	Leaves with stipules (often reduced to a ridge between the leaf-bases); corolla-lobes variously imbricate, or valvate in bud; plants usually woody	52
52a	(51) Corolla usually 5-lobed; stellate and/or glandular hairs absent	**180 Loganiaceae**
b	Corolla 4-lobed; stellate and glandular hairs present	**198 Buddlejaceae**

GROUP K

Flowers usually bisexual; calyx and corolla both present, the corolla made up of petals which are united, at least at the base, zygomorphic; ovary of 1 carpel or of several carpels.

1a Stamens more numerous than the corolla-lobes, or anthers opening by pores   2
  b Stamens as many as corolla-lobes or fewer, anthers not opening by pores   6
2a (1) Anthers opening by pores; leaves undivided; ovary of 2 or more united carpels   3
  b Anthers opening by longitudinal slits; leaves dissected or compound; ovary of a single carpel   5
3a (2) The 2 lateral sepals petal-like; filaments united
                                  103   **Polygalaceae**
  b No sepals petal-like; filaments free   4
4a (3) Shrubs with alternate or apparently whorled leaves; stamens 4–25   168   **Ericaceae**
  b Herbs with opposite leaves; stamens 5   182   **Gentianaceae**
5a (2) Leaves pinnate, or of 3 leaflets; perianth not spurred
                                  84   **Leguminosae**
  b Leaves laciniate; upper petals spurred; upper sepal helmet-like or spurred   41   **Ranunculaceae**
6a (1) Stamens as many as corolla-lobes; zygomorphy weak   7
  b Stamens fewer than corolla-lobes; zygomorphy pronounced   12
7a (6) Stamens on the same radii as the corolla-lobes; placentation free-central   173   **Primulaceae**
  b Stamens on different radii from the corolla-lobes; placentation axile   8
8a (7) Leaves of 3 leaflets, with translucent, aromatic glands; stamens 5, the upper 2 fertile, the lower 3 sterile   96   **Rutaceae**
  b Combination of characters not as above   9
9a (8) Ovary of 3 carpels; ovules many   187   **Polemoniaceae**
  b Ovary of 2 carpels; ovules 4 or many   10
10a (9) Flowers in coiled cymes; fruit of up to 4 1-seeded nutlets
                                  191   **Boraginaceae**

   b Flowers not in coiled cymes; fruit a many-seeded capsule   11

11a (10) Corolla-lobes contorted in bud; stamens 5, equal; leaves opposite; climber   **180 Loganiaceae**

   b Corolla-lobes various imbricate in bud; stamens 4 or 5 and unequal; leaves usually alternate   **199 Scrophulariaceae**

12a (6) Placentation axile; ovules 4 or many   13

   b Placentation parietal, free-central, apical or basal; ovules many or 1–2   20

13a (12) Ovules numerous but not in vertical rows in each cell of the ovary   14

   b Ovules 4, or more numerous but then in vertical rows in each cell of the ovary   16

14a (13) Seeds winged; mainly trees, shrubs or climbers with opposite, pinnate, digitate or rarely simple leaves   **201 Bignoniaceae**

   b Seeds usually wingless; mainly herbs or shrubs with simple leaves

     15

15a (14) Corolla-lobes imbricate in bud; septum of ovary in the horizontal plane; leaves opposite or alternate

     199 Scrophulariaceae

   b Corolla-lobes usually folded, valvate or contorted in bud; septum of ovary oblique, not in the horizontal plane; leaves alternate

     197 Solanaceae

16a (13) Leaves all alternate, usually with blackish, resinous glands; plants woody   **208 Myoporaceae**

   b At least the lower leaves opposite or whorled, none with glands as above; plants herbaceous or woody   17

17a (16) Fruit a capsule; ovules 4–many, usually in vertical rows in each cell of the ovary   18

   b Fruit not a capsule; ovules 4, side-by-side   19

18a (17) Leaves all opposite, often prominently marked with cystoliths; flower-stalks without swollen glands at the base; capsule usually opening elastically, seeds usually on hooked stalks

     202 Acanthaceae

   b Upper leaves alternate, cystoliths absent; flower-stalks with swollen glands at the base; capsule not elastic, seeds not on hooked stalks   **203 Pedaliaceae**

19a (17) Style arising from the depression between the 4 lobes of the ovary, or if terminal then corolla with a reduced upper lip; fruit usually of 4 1-seeded nutlets; calyx and corolla often 2-lipped

195 **Labiatae**

b Style terminal; corolla with well-developed upper lip; fruit usually a berry or drupe; calyx often more or less actinomorphic, not 2-lipped 193 **Verbenaceae**

20a (12) Ovules 4–many; fruit a capsule, rarely a berry or drupe 21

b Ovules 1–2; fruit indehiscent, often dispersed in the persistent calyx 27

21a (20) Ovary containing 4 ovules, side-by-side 193 **Verbenaceae**

b Ovary containing many ovules 22

22a (21) Placentation free-central; corolla spurred; leaves modified for trapping and digesting insects 207 **Lentibulariaceae**

b Placentation parietal or apical; corolla not spurred, rarely swollen at base; leaves not insectivorous 23

23a (22) Leaves scale-like, never green; root-parasites 24

b Leaves green, expanded; free-living plants 25

24a (23) Placentas 4; calyx laterally 2-lipped 206 **Orobanchaceae**

b Placentas 2; calyx 4-lobed 199 **Scrophulariaceae**

25a (23) Seeds winged; mainly climbers with opposite, pinnately divided leaves 201 **Bignoniaceae**

b Combination of characters not as above 26

26a (25) Capsule with a long beak separating into 2 curved horns; plant sticky-velvety 204 **Martyniaceae**

b Capsule without beak or horns; plant velvety or variously hairy or hairless 205 **Gesneriaceae**

27a (20) Flowers in heads surrounded by an involucre of bracts; ovule 1 200 **Globulariaceae**

b Flowers not in heads, often in spikes; ovules 1 or 2 28

28a (27) Fruits deflexed; calyx with hooked teeth; ovary 1-celled with 1 basal ovule 209 **Phrymaceae**

b Fruits mostly erect; calyx without hooks; ovary 2-celled, with a solitary apical ovule in each cell, fruit often 1-seeded

199 **Scrophulariaceae**

GROUP L

Monocotyledons with superior ovaries, or plants aquatic with totally submerged flowers.

1a Trees, shrubs or prickly scramblers with large, pleated, usually palmately or pinnately divided leaves; flowers more or less stalkless in fleshy spikes or panicles which often have large, sometimes woody basal bracts (spathes) **247 Palmae**

 b Combination of characters not as above   2

2a (1) Totally submerged aquatic plants of fresh or saline water   3

 b Terrestrial or epiphytic plants, if aquatic then not totally submerged, occasionally entirely floating   6

3a (2) Plants of fresh or brackish water; flowers bisexual, in axillary spikes, with a perianth of 4 segments which are valvate in bud, and 4 free carpels, *or* marine plants with densely fibrous rhizomes (often washed up on beaches as fibre-balls) with leaves mostly basal and bisexual flowers in stalked spikes subtended by reduced leaves   **226 Potamogetonaceae**

 b Combination of characters not matching either of the above   4

4a (3) Flowers on flattened axes or in 2-flowered spikes not enclosed by a leaf-sheath; plants of brackish or saline water

226 Potamogetonaceae

 b Flowers axillary, not enclosed by leaf-sheaths; plants of fresh or brackish water   5

5a (4) Leaves toothed; ovary apparently of a single carpel, though with 2–4 stigmas   **228 Najadaceae**

 b Leaves entire; ovary of 1–9 free carpels, the stigmas 1 per carpel, each dilated or 2–4-lobed   **227 Zanichelliaceae**

6a (2) Small, floating aquatic plants not differentiated into stem and leaves   **249 Lemnaceae**

 b Plants various, rarely floating aquatics, plant-body differentiated into stem and leaves   7

7a (6) Perianth entirely hyaline or papery or reduced to bristles, hairs, narrow scales or absent   8

 b Perianth well-developed though sometimes small, never entirely hyaline or papery   15

8a (7) Flowers in small, 2-sided or cylindric spikelets provided with overlapping bracts (spikelets sometimes 1-flowered)   9

  b Flowers arranged in heads, superposed spikes, racemes, panicles or cymes, never in spikelets as above   10

9a (8) Leaves alternate in 2 ranks on a stem which is usually hollow and with cylindric internodes; leaf-sheath usually with free margins, at least in the upper part; flowers arranged in 2-sided spikelets (sometimes 1-flowered) each usually subtended by 2 sterile bracts (glumes); each flower usually enclosed by a lower lemma and an upper palea (sometimes absent); perianth of 2–3 concealed scales (lodicules), more rarely of 6 scales or absent; styles generally 2, feathery   **246 Gramineae**

  b Leaves usually arranged on 3 sides of the cylindric or more usually 3-angled stems which usually have solid internodes; young leaf-sheaths closed, though sometimes splitting later; flowers arranged in 2-sided or cylindric spikelets, often with a 2-keeled or 2-lobed glume at the base; each flower subtended only by a glume; perianth of several bristles, hairs or scales, or absent; style 1 with 2 or 3 papillose stigmas   **253 Cyperaceae**

10a (8) Dioecious trees or shrubs often supported by stilt- roots, with stiffly leathery, sharply toothed leaves; fruits compound, often woody   **250 Pandanaceae**

  b Combination of characters not as above   11

11a (10) Inflorescence a simple fleshy spike (spadix) of inconspicuous flowers, subtended by or rarely joined to a large bract (spathe); leaves often net-veined, or lobed (plant rarely a small, evergreen, floating aquatic)   **248 Araceae**

  b Combination of characters not as above   12

12a (11) Flowers unisexual, in heads each surrounded by an involucre of bracts; perianth in 2 series, often greyish white

   245 Eriocaulaceae

  b Combination of characters not as above   13

13a (12) Flowers bisexual; perianth-segments 6, scarious or brownish; ovary with 3–many ovules   **240 Juncaceae**

  b Flowers unisexual; perianth-segments a few threads or scales; ovary with 1 ovule   14

14a (13) Flowers in 2 superposed, elongate, brownish or silvery spikes;
   ovary borne on a stalk with hair-like branches   252   **Typhaceae**
  b Flowers in spherical heads; ovary not stalked   251   **Sparganiaceae**
15a (7) Carpels free or slightly united at the base only   16
  b Carpels united for most of their length, though the styles may be
   free, or carpel solitary   20
16a (15) Inflorescence a spike, sometimes bifid; perianth-segments
   1—4   17
  b Inflorescence not a spike; perianth-segments 6   18
17a (16) Stamens 6 or more; carpels 3—6; perianth-segments 1—3,
   petal-like   224   **Aponogetonaceae**
  b Stamens 4; carpels 4; perianth-segments 4, not petal-like
                                           226   **Potamogetonaceae**
18a (16) Ovules many, borne on the walls of the free carpels
                                           221   **Butomaceae**
  b Ovules few, borne basally in each free carpel   19
19a (18) Leaf-sheaths with ligules; flowers in racemes;
   perianth-segments all similar, sepal-like   223   **Scheuchzeriaceae**
  b Leaf-sheaths without ligules; flowers in whorls, racemes or
   panicles; perianth differentiated into sepals and petals
                                           220   **Alismataceae**
20a (15) All perianth-segments similar   21
  b Perianth-segments of the outer and inner whorls conspicuously
   different, the former usually sepal-like, the latter petal-like 33
21a (20) Plants with scapes, with small flowers without bracts in
   racemes or spikes; perianth calyx-like; ovules 1 per cell, basal
                                           225   **Juncaginaceae**
  b Plants not as above; perianth usually corolla-like; ovules usually
   more than 1 per cell, rarely basal   22
22a (21) Inflorescence subtended by an entire, spathe-like sheath;
   plants aquatic   238   **Pontederiaceae**
  b Inflorescence not as above; plant terrestrial   23
23a (22) Perianth persistent into fruit, covered with branched hairs;
   stamens 3; sap usually orange   231   **Haemodoraceae**
  b Combination of characters not as above   24
24a (23) Aquatic herbs with submerged leaves which are 2-toothed at
   the apex   243   **Mayacaceae**

[86]

b Combination of characters not as above   25

25a (24) Outer perianth-whorl with 1 segment much larger than the
    others; segments of inner perianth-whorl petal-like, yellow; flowers
    in heads, with bracts   **244   Xyridaceae**

  b Combination of characters not as above   26

26a (25) Plant woody, or not woody but bearing rosettes of long-lived,
    fleshy or leathery leaves at or near ground level   27

  b Plant herbaceous, leaves usually not long-lived and in rosettes, if
    so then usually deciduous and not very fleshy   31

27a (26) Leaf-stalk bearing 2 tendrils; leaves net-veined   **229   Liliaceae**

  b Leaf-stalk without tendrils; leaves parallel-veined   28

28a (27) Leaves very small, scale-like or spiny, their functions taken
    over by flattened or thread-like stems (cladodes) on which the
    inflorescences are often borne   **229   Liliaceae**

  b Plant with expanded leaves, cladodes absent   29

29a (28) Shrubs or woody climbers with scattered stem-leaves; flowers
    solitary, usually large and hanging; placentation mostly parietal
                                                            229   Liliaceae

  b Combination of characters not as above   30

30a (29) Leaves leathery and more or less thin, if succulent then
    leathery or not, tip spine-like or cylindric; flowers usually
    greenish or whitish, campanulate or cup-shaped or with a narrow
    tube and spreading lobes, often more than 1 per bract
                                                            230 Agavaceae

  b Leaves succulent, usually without spine-like or cylindric tips;
    flowers usually red, yellow or orange, tubular, the lobes scarcely
    spreading, always 1 per bract   **229   Liliaceae**

31a (26) Leaves very small, scale-like or spiny, their function taken
    over by thread-like or flattened stems (cladodes) on which the
    inflorescences are often borne   **229   Liliaceae**

  b Plant with expanded leaves, cladodes absent   32

32a (31) Leaves evergreen, clearly stalked; flowers more than 1 to each
    bract, with a narrow tube as long as or longer than the spreading
    lobes   **230   Agavaceae**

  b Leaves deciduous, usually without distinct stalks; flowers of
    various shapes, rarely as above, always 1 per bract
                                                            229   Liliaceae

33a (20) Flowers solitary or in umbels; leaves broad, opposite or in a single whorl near the top of the stem **229 Liliaceae**

 b Flowers in spikes, heads, cymes or panicles; leaves not as above 34

34a (33) Stamens 6 or 3–5 with 1–3 staminodes; anthers basifixed; leaves usually borne on the stems, often with closedsheaths, never grey with scales; bracts neither overlapping nor conspicuously toothed **242 Commelinaceae**

 b Stamens 6, staminodes 0, anthers dorsifixed; leaves mostly in basal rosettes, often rigid and spiny-margined, when on the stems then usually grey with scales; bracts usually overlapping and conspicuously toothed **241 Bromeliaceae**

GROUP M

Monocotyledons with inferior or half-inferior ovaries.

1a Flowers actinomorphic or weakly zygomorphic; stamens 6, 4 or 3 or rarely many, very rarely 5, 2 or 1 2

 b Flowers strongly zygomorphic or asymmetric; stamens 5, 2 or 1, (very rarely 6), occasionally rather weakly zygomorphic, when stamen 1, united with the style to form a column 14

2a (1) Unisexual climbers with heart-shaped or very divided leaves; rootstock tuberous or woody **237 Dioscoreaceae**

 b Combination of characters not as above 3

3a (2) Perianth persistent into fruit, variously hairy; sap usually orange **231 Haemodoraceae**

 b Combination of characters not as above 4

4a (3) Rooted or floating aquatics; stamens 2–12; ovules distributed all over the carpel walls (diffuse-parietal placentation)

222 Hydrocharitaceae

 b Terrestrial or marsh plants, or epiphytes; stamens 3 or 6, rarely many; placentation axile or parietal (when ovules restricted to a few rows on carpel walls) 5

5a (4) Stamens 3, staminodes absent; leaves often sharply folded, their bases overlapping (equitant); style-branches often divided

239 Iridaceae

b Stamens 6, or 3 plus 3 staminodes; leaves usually not as above; style-branches not divided   6

6a (5) Placentation parietal; flowers in an umbel, with the inner bracts long, thread-like and hanging   236   **Taccaceae**

  b Placentation usually axile; inflorescence and bracts not as above   7

7a (6) Perianth consisting of an outer, calyx-like whorl and an inner, corolla-like whorl; bracts usually overlapping and conspicuously coloured   241   **Bromeliaceae**

  b Segments of perianth not in 2 dissimilar whorls as above; bracts not as above   8

8a (7) Ovary half-inferior   9

  b Ovary fully inferior   10

9a (8) Anthers opening by pores   233   **Tecophilaeaceae**

  b Anthers opening by slits   229   **Liliaceae**

10a (8) Leaves long-persistent, evergreen   11

  b Leaves dying down annually   12

11a (10) Leaves fleshy or leathery, thick, rigid or flexible, spine-tipped, often with spines or teeth or a removable stiff thread on the margins   230   **Agavaceae**

  b Leaves not spine-tipped, and without spines on their margins, somewhat leathery   235   **Velloziaceae**

12a (10) Flowers in a spike; leaves fleshy, often spotted with brown, the margins more or less rolled around each other in bud

   230   **Agavaceae**

  b Flowers in umbels or solitary; leaves not usually fleshy or spotted with brown, but flat, pleated or with the margins folded outwards in bud   13

13a (12) Leaves mostly basal, densely hairy, pleated or with prominent veins   234   **Hypoxidaceae**

  b Leaves various, not usually densely hairy, pleated or with prominent veins, basal or not   232   **Amaryllidaceae**

14a (1) Fertile stamens 6; perianth-segments all similar, united below into a curved and unevenly swollen tube; stem below ground, fleshy   230   **Agavaceae**

  b Stamens 5, 2 or 1, very rarely 6; staminodes, which may be petal-like, often present; perianth-segments usually differing among themselves; fleshy underground stems rare   15

15a (14) Fertile stamens 2 or 1, united with the style to form a
    column; pollen usually borne in coherent masses (pollinia);
    leaf-veins, when visible, all parallel to margins   **258   Orchidaceae**
  b Fertile stamens 5 or 1, rarely 6, not united with the style; pollen
    granular; leaf with a distinct midrib more or less parallel to
    margins, the secondary veins parallel to each other, running at an
    angle from midrib to margins   16
16a (15) Fertile stamens 5 or rarely 6   **254   Musaceae**
  b Fertile stamen 1, petal-like staminodes 5   17
17a (16) Fertile stamen with a thread-like filament and wider anther of
    2 pollen-bearing lobes, not petal-like   **255   Zingiberaceae**
  b Fertile stamen in part petal-like and with only 1 pollen- bearing
    anther-lobe   18
18a (17) Leaf-stalk with a swollen band (pulvinus) at the junction with
    the blade; ovary smooth, with 1–3 ovules   **257   Marantaceae**
  b Leaf-stalk without a pulvinus; ovary usually warty, with numerous
    ovules   **256   Cannaceae**

Arrangement and description of families

In general, the variation given in the descriptions is somewhat wider than that presented in the key, and many characters used in the latter have had to be omitted. Only the families keyed out in the main key have full entries here, but some segregate families are also keyed out and briefly described under the relevant major families.

No attempt has been made to diagnose the orders in which families are grouped, but for each a few differential features are cited, most of which are repeated in the descriptions of the families. The circumscription and content of the orders remain largely matters of opinion.

In using the descriptions, the following features must be assumed for most species of a family, unless otherwise stated: milky sap absent, habit not succulent, parts of the flower free from each other, stamens not antepetalous and anthers opening by longitudinal slits. The ptyxis, as far as it is known, is given for each family; this refers to leaves if they are undivided, to leaflets if the leaves are divided.

The following points concerned with presentation should be noted.

Morphology

The oblique stroke (/) is used instead of 'or'; the letter 'n' is used instead of 'many' (i.e. more than 10 or 12). K – number of calyx-segments, C – number of corolla-segments, P – number of perianth-segments when these are undifferentiated, A – number of stamens, G – number of carpels. These letters are also used in the collective sense, e.g. 'A antepetalous' means stamens antepetalous. Brackets are used to indicate that the segments of any particular whorl are united to each other, e.g. C(5) means a corolla of 5 united segments (lobes).

In the Dicotyledons we have not usually indicated whether the ovary is superior or inferior; however, we have always stated whether or not the perianth and stamens are hypogynous, perigynous or epigynous, which gives the same information (see pp. 32–41). Many of the families with the petals united into a tube at the base are described as 'K hypogynous, CA perigynous'; this means that the stamens are borne on the corolla and the ovary is superior. In the Monocotyledons, the position of the ovary is always indicated.

In plants with inferior ovaries, the number of the calyx-segments is shown as free (i.e. unbracketed) if the segments are completely free above their point of attachment on the ovary, and united if they are united above this point. Accurate information about this feature is difficult to obtain.

Information about the inflorescence-type is always given, but this, again, is difficult to obtain and condense; the information given here should not be regarded as a complete description of the range of inflorescence-types found in any particular family. We have referred to six main types of inflorescence: racemes, spikes, panicles, cymes, umbels and heads; the terms 'clusters' or 'fascicles' have also been used when the precise nature of the inflorescence is not easily understood. 'Racemose' has been used when spikes, racemes or basically indeterminate panicles occur in the same family; 'cymose' has been used in a similar way to cover various types of determinate inflorescences.

Ovule number refers to the number of ovules in each free carpel (if the ovary is made up of free carpels) or to the ovary as a whole (when it is made up of united carpels), unless qualified by 'per cell'.

Geography

This is indicated in two main ways. At the end of each family description there is a summary of its total distribution; most of the terms used for this are self-explanatory. In the brief paragraph of observations following each family description, the number of genera occurring in Europe and/or North America is given; these figures are derived from Tutin *et al.*, *Flora Europaea*, volumes 1–5 (1964–1980) for Europe, and from Kartesz, J. T. & R., *A synonymised Checklist of the Vascular Flora of the United States, Canada and Greenland* (1980) for North America. Some details of the number of genera in each family that are important in amenity horticulture in Europe are also given.

General

For monogeneric families the name of the single genus is given in the observations to the family. Some synonyms and short comments about the relationships of some families are also included.

The permitted alternative family names (for the eight families whose traditionally used names do not end in '-aceae') are given, separated from the more familiar name by an oblique stroke, e.g. 164. **Umbelliferae/Apiaceae**.

Subclass Dicotyledones

Cotyledons usually 2, lateral; leaves usually net-veined, with or without stipules, alternate, opposite or whorled; flowers with parts in 2s, 4s or 5s, or parts numerous; primary root-system (taproot) usually persistent, branched.

CASUARINALES

Woody, monoecious; branches whip-like; leaves reduced.

1 **Casuarinaceae**. Woody, branches jointed. Leaves whorled, scale-like. Inflorescence a catkin. Flowers unisexual, P0 (?) A1, G(2), naked; ovules 2, parietal; styles 2, free. Samaras in woody cones. *Australasia to Malaysia*.

A single genus (*Casuarina*) of 70 species with a few of them in cultivation as glasshouse shrubs.

JUGLANDALES

Monoecious; leaves often pinnately compound; inflorescence a catkin.

2 Myricaceae. Woody. Leaves alternate, entire/divided, exstipulate, aromatic, gland-dotted; ptyxis conduplicate. Inflorescence a catkin. Flowers usually unisexual. P0, A2–20 usually 4–8, G(2), naked; ovule 1, basal; styles 2, free. Drupe. *N hemisphere.*

A family of 3 genera and about 30 species; 2 genera native in North America, 1 native in Europe. A few species cultivated as ornamental, aromatic shrubs.

3 Juglandaceae. Woody. Leaves usually opposite, pinnately compound, exstipulate; ptyxis conduplicate. At least the male flowers in catkins. Flowers unisexual. P4, A3–n, G(2–3), inferior; ovule 1, basal; styles 2, free, sometimes divided, stigmas internal, lateral. Nut with complex lobed and folded cotyledons. *N temperate areas.*

Seven genera and about 40 species. Two genera are native in Europe (with 1 introduced) and 2 are native in North America. Four genera are cultivated as ornamental trees, and species of *Juglans* (walnut) and *Carya* (pecan) are cultivated for their edible nuts.

LEITNERIALES

Dioecious shrubs, catkinate; fruit a drupe.

4 Leitneriaceae. Woody. Leaves alternate, entire, exstipulate. Inflorescence a catkin. Flowers variously interpreted; male: P0, A3–12; female: P3–8, G1-celled, superior; ovule 1, lateral; style 1, stigma lateral. Drupe. *USA.*

The 1 genus, *Leitneria*, has only a single species; it occurs in the USA and is not or rarely cultivated in Europe.

SALICALES

Dioecious, woody; ovules parietal; fruit a capsule, seeds woolly.

5 **Salicaceae.** Woody, dioecious. Leaves usually alternate, simple, stipulate; ptyxis involute (*Populus*) or supervolute (*Salix*). Inflorescence a catkin. Flowers usually unisexual with disc or nectary-gland. P0 A2–n/(2–n), G(2–4), superior; ovules n, parietal; stigmas 2–4 on a short style or sessile. Capsule; seeds woolly. *Widespread.*

Two genera, *Salix* with more than 300 species and *Populus* with about 40 species. Both genera are native to North America and Europe, and species of both are cultivated as ornamental trees and shrubs.

FAGALES

Woody, monoecious; leaves stipulate; at least the male flowers in catkins; G inferior/naked; fruit a nut.

6 **Betulaceae.** Woody. Leaves alternate, simple, stipulate; ptyxis conduplicate-plicate. At least the male flowers in catkins. Flowers unisexual: male: P(4)/0, A2/4/2–20; female: P0/irregularly lobed, G(2), naked/inferior; ovule 1 per cell, axile/apical; styles 2, free. Nut, either small, often winged and in 'cones'/larger, clasped in bract/cupule. *N temperate areas.*

Six genera and about 150 species. Often divided into 2 families.

1a Nuts small, borne in 'cones'; perianth present in male flowers, absent in female; ovary naked   **6a Betulaceae.**
 b Nuts larger, subtended by leaf-like bracts or involucres (cupules); perianth present in female flowers, absent in male; ovary inferior
            6b Corylaceae

6a **Betulaceae** in the strict sense. Trees or rarely small shrubs. Male flowers: in pendent catkins, generally 3 to a bract, P4, A2/4; female flowers: in catkins or erect 'cones', perianth absent, ovary naked. *Mostly N temperate areas.*

Two genera, *Betula* (60 species) and *Alnus* (15 species) both native in Europe and North America, and both widely cultivated as ornamental trees and shrubs.

6b Corylaceae. Trees or small shrubs. Male flowers usually in pendulous catkins, P0, A4–15; female flowers: in catkins or short spikes, P irregularly lobed. *N temperate areas.*

There are 4 genera, of which 3 are native in Europe and 3 in North America. Species of all 4 genera are grown as ornamentals.

7 Fagaceae. Woody; leaves usually alternate, simple, stipulate; ptyxis conduplicate, conduplicate-plicate or supervolute (important in the classification of *Nothofagus*). Male inflorescence a catkin. Flowers unisexual: male: P(4–7), A4–n; female: P(4–7). G(3–6), inferior; ovules 2 per cell, axile; styles 3–6, free. Nut enveloped in cupule. *Temperate & tropical areas.*

Eight genera and about 600 species. Three genera native to Europe, 5 to North America. Species of many genera are important timber trees, and those of 7 genera are cultivated as ornamentals.

URTICALES

Leaves simple, alternate; flowers small, often unisexual, without petals; ovary usually superior; ovule 1.

8 Ulmaceae. Woody. Leaves alternate, simple, stipulate, usually with oblique bases; ptyxis conduplicate. Flowers solitary/clustered, unisexual/bisexual, zygomorphic. PA hypogynous. P(4–9), A4–9, G(2); ovule 1, apical; styles 2, free, sometimes divided above. Samara/drupe. *N hemisphere.*

A family of 16 genera and about 140 species. Three genera native to Europe, 5 to North America. Species of 3 genera are grown as ornamental trees or shrubs.

9 Eucommiaceae. Trees with milky sap. Leaves alternate, simple, exstipulate; ptyxis supervolute. Flowers solitary, unisexual, actinomorphic. P0, A4–10, G(2) naked; ovule 1, apical; styles 2, free. Samara. *China*.

A family of a single genus and species (*Eucommia ulmoides*). It is grown as an ornamental and has some interest in that its milky sap contains rubber.

10   **Moraceae**. Woody/herbs/climbers, often with milky sap. Leaves alternate/opposite, simple/divided, stipulate; ptyxis conduplicate or supervolute (important in the classification of *Ficus*). Inflorescence various, flowers sometimes sunk in expanded receptacle, which may take the form of a hollow cup. Flowers unisexual. PA hypogynous. Male: P2–6 usually 4/(5); A1–5; female: P2–6 or entire and enveloping ovary, G(2), often 1 carpel aborting; ovule 1, apical; styles 1–2, free. Achene/syncarps. *Tropical & N temperate areas*.

There are about 50 genera and up to 1,500 species. Often divided into 2 families.

1a Usually woody plants with milky sap; perianth in male flowers of
   2–6 free segments; fruit usually a syncarp   10a   **Moraceae**
 b Herbs without milky sap; perianth in male flowers of 5 united
   segments; fruit an achene   10b   **Cannabaceae**

10a   **Moraceae** in the strict sense. Plants woody with milky sap. Perianth in male flowers usually of 2–6 free segments. Fruit usually a syncarp. *Mainly tropics*.

As for the family in the broad sense, there are about 50 genera and up to 1,500 species. Four genera (3 of them introduced) occur in Europe and 17 genera in North America. Species from 7 genera are cultivated as ornamentals. Species of *Ficus* (figs) and *Artocarpus* (breadfruit) are economically important.

10b   **Cannabaceae**. Herbs or non-woody climbers without milky sap. Perianth in male flowers (5); fruit an achene. *Temperate Eurasia*.

Two genera, *Humulus* and *Cannabis*. *Humulus* (hops) is native to Europe, while *Cannabis*, now widely cultivated all over the world often (illegally) for its narcotic resin or legally for fibre, probably originated in Central Asia.

11 **Urticaceae.** Herbs/rarely shrubs, often with rough or stinging hairs. Leaves alternate/opposite, simple, usually stipulate; ptyxis conduplicate or involute. Inflorescence various. Flowers unisexual, actinomorphic. PA usually hypogynous/ovary naked. P0–5/(2–5), A3–5 usually 4 inflexed in bud and touch-sensitive, G1; ovule 1, basal; style 1/rarely 0, stigma often brush-like. Achene/drupe. *Widespread.*

There are about 50 genera and 2,000 species. Five genera occur in Europe and 13 in North America. Species of several genera are cultivated as fibre-plants, and of about 8 genera as ornamentals.

PROTEALES

Woody, apetalous; P usually (4); G1.

12 **Proteaceae.** Trees/shrubs. Leaves alternate, exstipulate, evergreen, often very hard; ptyxis flat or conduplicate. Inflorescence various. Flowers usually bisexual, actinomorphic/zygomorphic. PA perigynous. P(4), A4 rarely 1–3 infertile, borne on petal-like, spoonshaped P-segments, G1; ovules 1–n, marginal; style 1, thickened or with pollen-collecting apparatus at the apex. Follicle/nut/drupe. *S hemisphere.*

There are thought to be about 75 genera and up to 1,300 species. Many are highly ornamental, though difficult to grow; species of about 16 genera are in cultivation.

SANTALALES

Often parasitic or partially so, usually green; C absent; G inferior.

13 **Olacaceae.** Woody, sometimes half-parasitic on the roots of other trees. Leaves alternate, simple, exstipulate. Flowers solitary/in clusters/racemes/panicles, bisexual, actinomorphic. KCA hypogynous. K4–6, C3–5/(6), A3+5 staminodes/5/8–10, G(3) usually

1-celled; ovules 2–3, axile; style 1, stigma 2–3-lobed. Drupe. *Mostly tropics*.

A family of 27 genera and about 320 species. Two genera occur in North America. A few species are occasionally cultivated as ornamentals or curiosities.

14 **Santalaceae**. Herbs/shrubs/trees, parasitic on the roots of other plants. Leaves alternate/opposite, simple, exstipulate. Inflorescence spike/raceme/cluster/flowers solitary. Flowers unisexual/bisexual, actinomorphic. PA epigynous/rarely half-epigynous. P3–5/(3–5), A3–5, G(3–5); ovules 1–5, basal; style 1, stigma 2–5-lobed. Nut/drupe. *Widespread*.

Thirty-five genera with about 400 species; 3 genera occur in Europe, 8 in North America. Species of perhaps 6 genera are occasionally grown as ornamentals.

15 **Loranthaceae**. Mostly woody/herbaceous branch-parasites. Leaves opposite/whorled, simple, exstipulate, usually evergreen. Flowers unisexual, actinomorphic, solitary/in pairs. PA epigynous. P4–6/(4–6), A2–6, G(3–6); ovules not differentiated in flower; style absent, stigma sessile, unlobed. Berry/drupe, 2–3-seeded. *Mostly tropics*.

There are 40 genera and about 1,400 species, almost all of them parasitic. Three genera occur in Europe, 8 in North America. Species of *Viscum* and *Loranthus* are occasionally cultivated; *Viscum album* (mistletoe) is an important ornamental used as a Christmas decoration.

POLYGONALES

Stipules often prominent and united into a sheath; ovule 1, basal.

16 **Polygonaceae**. Herbs/shrubs/climbers/rarely trees. Leaves alternate or all basal, simple/lobed, stipules usually present, often united into a sheath (ochrea); ptyxis revolute. Flowers in racemes/cymes, unisexual/bisexual, actinomorphic. PA usually hypogynous. P3–6/rarely (3–6), A6–9, G(2–4), usually (3); ovule 1, basal; styles 2–4,

free/occasionally slightly united at base. Nut. *Mainly N temperate areas*.

There are about 30 genera with some 800 species; 12 genera occur in Europe and 24 in North America. Species of about 10 genera are cultivated as ornamentals.

CARYOPHYLLALES

C free/0; embryo usually strongly curved round perisperm.

17 **Phytolaccaceae**. Trees/shrubs (some climbing)/herbs. Leaves alternate, entire, exstipulate; ptyxis conduplicate. Flowers in racemes, usually bisexual, actinomorphic. PA hypogynous. P4–5/(4–5), A3–n, G1–(n)/rarely n; ovules 1/1 per cell, basal/axile. Fruit often fleshy. *Mainly American tropics & S hemisphere*.

There are about 20 genera and 100 species. One genus is introduced into Europe, but 8 are native in North America. Species of 4 genera are commonly cultivated as ornamentals.

18 **Nyctaginaceae**. Herbs/shrubs/woody climbers. Leaves alternate/opposite/whorled, usually opposite, entire, exstipulate; ptyxis conduplicate. Inflorescence cymose. Flowers usually bisexual, actinomorphic. PA hypogynous. P(5), A1–n/(1–n) usually 5/(5), G1-celled; ovule 1, basal; style 1, stigmas slightly divided at apex. Achene, often in persistent P. *Mostly tropics*.

There are 30 genera and about 300 species; 3 genera (2 introduced) occur in Europe, whereas 16 genera are native in North America. Species and hybrids of a few genera, notably *Mirabilis* (Marvel of Peru) and *Bougainvillea* are widely grown as ornamentals.

19 **Aizoaceae**. Herbs/shrubs, usually leaf succulents. Leaves usually opposite, often all basal, simple, exstipulate. Flowers in cymes, bisexual, actinomorphic. KCA usually epigynous/rarely half epigynous. K4–15/(4–15), C n, A n, G(3–20); ovules usually n, parietal/rarely axile; styles free or stigmas sessile, radiating. Fruit usually a complex capsule opening when wet, closing when dry/rarely fleshy and indehiscent. *Mostly South Africa*.

There are about 120 genera and some 2,500 species, mostly from South Africa and adjacent areas; most of the genera were originally part of the large genus *Mesembryanthemum*, but taxonomic work has shown their distinctiveness. Species of many genera are grown in cultivation as ornamental succulents.

The family sometimes separated off as the *Molluginaceae* is included here.

20 **Portulacaceae**. Herbs/shrubs, often fleshy. Leaves alternate/opposite, simple, entire, stipulate/exstipulate; ptyxis very variable. Inflorescence racemes/cymes/rarely flowers solitary. Flowers bisexual, actinomorphic. KCA hypogynous/partially epigynous. K2/rarely 3 or more, C3–18/(3–18) when united only so at extreme base, usually 4–6, A3–n on the same radii as petals when few, G(2–8), 1-celled; ovules 1–n, basal/free central; styles 2–8, very slightly united below/rarely single. Capsule/rarely indehiscent. *Mostly New World*.

Nineteen genera with about 500 species. Two genera are native to Europe and 8 to North America. Species of 9 genera are cultivated as ornamentals and *Portulaca oleracea* is widely grown as a salad.

21 **Basellaceae**. Climbers. Leaves alternate, simple, fleshy, exstipulate. Inflorescence racemose. Flowers unisexual/bisexual, actinomorphic. PA perigynous. P5/(5), A5, G(3), 1-celled; ovule 1, basal; style 1 stigmas usually 3. Drupe in persistent, fleshy P. *Mostly tropical America*.

There are 4 genera and about 17 species. Two genera occur in North America, and a few species are cultivated for ornament.

22 **Caryophyllaceae**. Herbs/rarely shrublets. Leaves usually opposite, simple, entire, exstipulate/rarely stipulate; ptyxis flat, conduplicate or rarely supervolute. Inflorescence cymose/flowers solitary. Flowers usually bisexual, actinomorphic. KCA hypogynous/rarely perigynous/rarely PA hypogynous/perigynous. K4–5/(4–5), C4–5/rarely 0, A8–10/rarely fewer, G(2–5); ovules usually n, free-central/rarely 1 basal; styles 2–5, free/rarely 1. Capsule/nut/rarely fleshy. *Mostly N temperate areas*.

A family of about 90 genera and more than 2,000 species. Thirty-seven genera are native in Europe, and 35 in North America. The family is one of the most important in providing ornamentals, especially species of *Dianthus* (pinks, carnations), *Saponaria* (soapwort), *Silene*, *Gypsophila*, etc.; species of about 20 genera are found in European gardens.

23 Chenopodiaceae. Herbs/shrubs, often succulent. Leaves alternate/opposite, usually simple, exstipulate, reduced to scales when stems fleshy and segmented. Inflorescence usually cymose. Flowers unisexual/bisexual, actinomorphic. PA usually hypogynous. P(3–5)/rarely 0, green/membranous, A usually 5, G(2–3), 1-celled, rarely half-inferior; ovule 1, basal; styles 2–5, usually free. Achene/nut. *Widespread*.

About 100 genera and 1,400 species. Thirty-four genera occur in Europe and 25 in North America. Very few are cultivated, though species from about 10 genera can be found in European gardens, mostly grown as foliage plants.

24 Amaranthaceae. Herbs/woody. Leaves alternate/opposite/whorled, usually entire, exstipulate; ptyxis flat or conduplicate. Inflorescence often racemose and very condensed. Flowers usually bisexual, actinomorphic. PA hypogynous. P3–5/(3–5) usually hyaline and/or papery, A usually (5) staminodes frequent, G(2–3); ovules 1–n, basal; style 1, stigma slightly lobed/styles 2–3. Capsule/achene/berry. *Mostly tropics*.

Sixty genera and 900 species are known, 4 genera occurring in Europe, 18 in North America. A few (especially species of *Amaranthus*) are in cultivation.

25 Didieriaceae. Woody, succulent, often spiny. Leaves alternate, exstipulate. Cymes/panicles/umbels. Flowers unisexual, actinomorphic. KCA hypogynous. K2 persistent, C4 in 2 pairs, A usually 8/10, G(3–4), 1-celled; ovule 1, basal; style 1, stigma 3–4-lobed. Fruit dry, dehiscent. *Madagascar*.

Four genera and 11 species; a few are grown as curiosities.

CACTALES

Succulents, often spiny; leaves usually 0; K & C n.

26 **Cactaceae**. Mostly spiny stem-succulents. Leaves usually absent, when present, ptyxis supervolute (*Pereskia*). Flowers usually solitary and bisexual, actinomorphic/slightly zygomorphic. KCA usually epigynous. K n, C n/(n), A n, G(3–n); ovules n, parietal; style 1, stigma much-divided at apex. Berry. *Mostly America.*

There are over 100 genera and at least 1,500 species, all from the New World except for 1 which occurs in Africa, Madagascar and Sri Lanka. Nineteen genera occur in North America. Material of many genera and species is cultivated.

MAGNOLIALES

Aromatic trees/shrubs/climbers; G1–n, carpels usually free.

27 **Magnoliaceae**. Trees/shrubs. Leaves simple, deciduous/evergreen, alternate with large, deciduous stipules which enclose buds; ptyxis conduplicate. Flowers bisexual, actinomorphic, solitary. PA/KCA hypogynous. P/KC in several series, in 3s or rarely 4s (usually 6 or 9), A n spirally arranged, G n spirally arranged; ovules 2–n, marginal; styles 1 per carpel. Fruit a group of follicles; seeds large. *Mostly N temperate & subtropical areas.*

A family of 12 genera and about 200 species. Two genera are native in North America, and species of 3 genera are widely cultivated (especially species of *Magnolia*).

28 **Winteraceae**. Woody. Leaves alternate, simple, evergreen, exstipulate; ptyxis supervolute (*Drimys*). Flowers unisexual/bisexual, actinomorphic, in cymes/fascicles. KCA hypogynous. K2–6/(2–6) valvate, C2–n in 2–several series, A n, G1–n in 1 whorl; ovules 1–n, marginal; styles 1 per carpel, free. Follicle/berry. *Tropics (except Africa), S temperate areas.*

Possibly 6 genera and about 80 species. A few species of *Drimys* and *Pseudowintera* are cultivated as ornamental shrubs.

[103]

29 **Annonaceae**. Woody. Leaves simple, deciduous/evergreen, exstipulate; ptyxis conduplicate. Inflorescence various. Flowers usually bisexual, actinomorphic. KCA hypogynous. K usually 3, C3–6, K & C sometimes not distinguishable, A n each crowned by an enlarged connective, G n usually stalked in fruit; rarely united into a mass; ovules 1–n, basal/marginal; styles 1 per carpel, free. Berry or aggregate of berries; seeds with arils, endosperm convoluted. *Tropics, N temperate areas in New World.*

A large family of 120 genera and about 2,100 species, mainly tropical, but with 6 genera native in North America. Species of about 6 genera are cultivated as ornamentals.

30 **Myristicaceae**. Dioecious trees. Leaves alternate, entire, evergreen, exstipulate. Inflorescence racemose. Flowers unisexual, actinomorphic. PA hypogynous. P(2–5) usually (3), valvate in bud; male: A(2–10) filaments united; female: G1; ovule 1, basal; style 1. Fruit fleshy, 2-valved; seed with coloured aril, endosperm convoluted. *Tropics.*

There are 15 genera and about 250 species. Species of *Myristica* and *Pycnanthus* are occasionally found in cultivation (in glasshouses) and *Myristica fragrans* is the source of nutmeg and mace (seed and aril, respectively).

31 **Canellaceae**. Woody, bark very aromatic. Leaves alternate, simple, exstipulate. Inflorescence cymes/racemes. Flowers bisexual, actinomorphic. KCA hypogynous. K3, C4–5, A (more than 10) filaments united, G(2–6), 1-celled; ovules 2–n, parietal; style 1, short, stigma 2–6-lobed. Berry. *Tropical & subtropical Africa & America.*

A family of 6 genera and about 20 species; 2 genera are native to North America. Material of *Canella* is occasionally grown.

32 **Schisandraceae**. Woody climbers, monoecious or dioecious. Leaves alternate, simple, exstipulate; ptyxis involute (*Kadsura*) or supervolute (*Schisandra*). Flowers unisexual, actinomorphic, axillary. PA hypogynous. P5–20, occasionally poorly differentiated into K- and C-like whorls, A(n) usually united into a fleshy mass, G n;

ovules 2–3, marginal; styles free, 1 per carpel. Fruit berry-like, crowded or distant on an elongate axis. *N America, E Asia.*

A family of 2 genera and about 45 species. One genus (*Schisandra*) is native in North America, and species of *Schisandra* and *Kadsura* are grown as ornamental climbers.

33　Illiciaceae. Woody, aromatic. Leaves alternate/whorled, evergreen, simple, exstipulate; ptyxis supervolute. Flowers bisexual, actinomorphic, solitary. PA hypogynous. P7–n, in several whorls, imbricate, occasionally distinguishable by size into K- and C-like whorls, A4–n. G5–n in a single whorl; ovule 1, almost basal; styles 1 per carpel, free. Fruit a group of follicles. *SE Asia, SE North America.*

A family of a single genus (*Illicium*) with about 40 species; 2 of the species are native to North America and several are cultivated as aromatic ornamental shrubs.

34　**Monimiaceae.** Trees/shrubs, aromatic. Leaves opposite, simple, exstipulate, usually evergreen; ptyxis conduplicate. Flowers solitary/cymose, usually unisexual, actinomorphic. PA perigynous. P4–n occasionally in 2 whorls but all similar, A usually n each with a pair of glands or cup-like appendage at base, G1–n; ovule 1, basal; styles free, 1 per carpel. Achenes in enlarged perigynous cup. *Mainly tropics.*

A family of 34 genera and about 450 species. A few species of *Atherosperma, Laurelia* and *Peumus* are grown as ornamental trees and shrubs in glasshouses.

35　**Calycanthaceae.** Shrubs, aromatic. Leaves opposite, entire, deciduous/rarely evergreen, exstipulate; ptyxis flat-conduplicate. Flowers solitary, bisexual, actinomorphic. PA perigynous. P n, A5–30, G n; ovules 1–2, marginal; styles 1 per carpel, free. Achenes in persistent perigynous zone. *China, SE USA.*

Two genera with 9–10 species. One genus (*Calycanthus*) is native to North America, and species of it and of *Chimonanthus* (China) are cultivated as early-flowering shrubs.

[105]

36 **Lauraceae**. Woody, aromatic. Leaves usually alternate, entire, exstipulate, usually evergreen, glandular-punctate; ptyxis conduplicate or supervolute. Inflorescence cymose/racemose. Flowers small, unisexual/bisexual, actinomorphic. PA hypogynous/perigynous. P usually (4/6), A8–12/variable, anthers opening by valves, G1; ovule 1, apical; style 1. Drupe-like berry. *Mostly tropics*.

There are nearly 50 genera and about 2,000 species, mostly tropical, but with 2 genera occurring in Europe (1 introduced) and 13 native in North America. Species of a few genera are grown as ornamental shrubs. *Laurus nobilis* is the bay, used as a flavouring, and *Persea indica* produces the Avocado pear.

37 **Tetracentraceae**. Woody. Leaves alternate, simple, exstipulate. Inflorescence catkin-like. Flowers bisexual, actinomorphic. PA hypogynous. P4, A4, G(4); ovules n, apical; styles 4, free. Capsule. *China & adjacent Burma*.

A family of a single genus (*Tetracentron*) and species, now commonly grown in European gardens.

38 **Trochodendraceae**. Woody. Leaves whorled, simple, exstipulate, evergreen; ptyxis supervolute. Flowers actinomorphic, bisexual, racemose/in clusters. PA hypogynous. P minute/0, A n, G(6–10) in a single whorl; ovules n, marginal; styles 6–10, free. Fruit a group of coalesced follicles. *Japan*.

Like the *Tetracentraceae*, a family of a single genus (*Trochodendron*) and species which is widely cultivated.

39 **Eupteleaceae**. Woody. Leaves alternate but in false whorls, deciduous, exstipulate; ptyxis conduplicate. Flowers actinomorphic, bisexual, in the leaf-axils. P0, A n hypogynous, G6–n, stalked; ovules 1–3, marginal; styles 6–n, free. Fruit a group of stalked samaras. *Himalaya, Japan, China*.

A family of 1 genus (*Euptelea*) and 2 very similar species, occasionally grown as ornamental small trees.

40 **Cercidiphyllaceae**. Trees, dioecious. Leaves deciduous, simple, opposite on long shoots, alternate on short shoots, stipulate; ptyxis

involute. Flowers unisexual; male: almost stalkless, P4 A15–20; female: stalked, P4 hypogynous, G4–6; ovules n, marginal; styles 4–6, free. Follicles. *China, Japan.*

One genus (*Cercidiphyllum*) with a single species (occasionally regarded as 2 species) cultivated as a handsome shrub or small tree.

41 **Ranunculaceae.** Usually herbs/rarely woody/rarely climbing. Leaves usually alternate, simple/compound, usually exstipulate; ptyxis variable, mainly conduplicate or supervolute. Inflorescence various. Flowers bisexual, actinomorphic/zygomorphic. KCA/PA hypogynous. P4–n/K3–5, C2–n/rarely (4), bearing nectaries, A5–10/n, G1–n/rarely united; ovules 1–n, marginal/basal/apical, rarely axile; styles 1 per carpel, long or short, free. Achenes/follicles/rarely berry-like. *Mainly temperate areas.*

A family of about 50 genera and over 2,000 species. Twenty-three genera are native in Europe and 24 in North America. The family provides many ornamental herbaceous plants, especially in the genera *Aconitum, Anemone, Aquilegia, Delphinium, Helleborus, Pulsatilla* and *Ranunculus.*

42 **Berberidaceae.** Herbs/shrubs. Leaves alternate/rarely opposite, simple/divided, usually exstipulate, evergreen/deciduous; ptyxis variable. Inflorescence cymose/racemose/flowers solitary. Flowers bisexual, actinomorphic. KCA hypogynous, K & C sometimes not well differentiated. K3–n/rarely 0, C4–6/rarely 9/rarely 0, A4–18, often on the same radii as the petals, anthers often opening by valves, G apparently 1; ovules few, basal/marginal; style 1, short. Capsule/berry. *Mainly N temperate areas.*

Sometimes divided into several families, including *Berberidaceae* in the narrow sense, *Nandinaceae, Diphyllaeiaceae, Leonticaceae* and *Podophyllaceae.*

In the broad sense, there are 15 genera and 570 species; 6 genera (1 introduced) are found in Europe and 10 in North America. Many genera, notably *Berberis, Epimedium* and *Mahonia* are very ornamental, and are widely cultivated.

43   **Lardizabalaceae.** Woody, monoecious or dioecious, usually climbers. Leaves alternate, compound, exstipulate; ptyxis conduplicate. Inflorescence raceme-like. Flowers usually unisexual, actinomorphic. PA hypogynous. P6, sometimes in 2 whorls/K3–6, C6, A6/(6)/rarely 0, G3–10; ovules n, marginal; styles 1 per carpel, stigmas sessile. Berries. *Scattered.*

A family of 8 genera and 21 species, 1 genus native in North America. Species of about 5 genera are found in cultivation, as ornamental climbers.

44   **Menispermaceae.** Usually woody climbers. Leaves alternate, simple, deciduous/evergreen, exstipulate; ptyxis flat or conduplicate. Flowers unisexual, actinomorphic, in racemes/panicles. KCA/PA hypogynous. K4–n, C6–9/P3/6/9, A3/6/9/n, G3–6; ovules 1 per carpel, marginal; styles 3–6, free. Drupe/achene; seeds weakly to conspicuously horseshoe-shaped. *Mostly tropics.*

There are 67 genera and 470 species. Five genera are native to North America. Species of some 7 genera are grown as ornamental climbers.

45   **Nymphaeaceae.** Rhizomatous aquatic herbs. Leaves alternate, cordate/peltate; ptyxis involute. Flowers bisexual, actinomorphic, solitary. PA/KCA hypogynous or serially attached to ovary. P n/K3–6, C3–n, A n, G3–n/(3–n); ovules few–n on walls of carpels (parietal); styles 1 per carpel or stigmas sessile. Fruit various. *Widespread.*

There are about 9 genera with 100 species. Two genera are native in Europe, and 5 in North America. It is often divided into several smaller families.

1a	Carpels free	2
b	Carpels more or less united	3
2a	Carpels sunk individually in a top-shaped receptacle; perianth-segments in several series	**45d Nelumbonaceae**
b	Carpels not sunk in the receptacle; perianth-segments in 2 series	**45c. Cabombaceae**
3a	Ovary inferior; leaves with prickles	**45b Euryalaceae**

45a   Nymphaeaceae in the strict sense. Leaves and flowers usually floating, rarely somewhat raised above the water-surface. Leaves all alternate or basal, with cordate bases with conspicuous sinuses, without prickles. K4–6, C n, A n, G(3–35), superior; ovules n per cell. *Widespread.*

Nymphaea and *Nuphar* are native in both Europe and North America.

45b   Euryalaceae. Very similar to *Nymphaeaceae* in the strict sense, but often very large, leaves peltate, without conspicuous sinuses, with prickles all over the lower surface. G(8–n), inferior. *Scattered (India & E Asia, S America).*

45c   Cabombaceae. Leaves both submerged and floating, the sub-merged leaves opposite or whorled, very finely divided, the floating leaves entire (occasionally absent), peltate, without sinuses. Flowers solitary, floating or raised above the water-surface. K3, C3, A3–6, G2–4, superior; ovules usually 3 per carpel, pendulous. *Scattered (N & S America, Africa, NE Asia).*

Cabomba is native to Atlantic North America.

45d   Nelumbonaceae. Leaves alternate, floating when young, held well above the water when mature, peltate, widely funnel-shaped, without sinuses. Flowers held well above the water. K4–5, C10–25, A n, each appendaged, G12–20, free and sunk individually in a top-shaped receptacle; ovule 1 per carpel, apical. *Warm temperate and tropical America & Asia.*

Nelumbo is the sole genus. *N. lutea* is native to the eastern and southern parts of the USA.

46   Ceratophyllaceae. Submerged aquatic herbs. Leaves whorled, much divided, exstipulate. Flowers solitary, unisexual, actino-morphic. PA hypogynous. P(8–13), A10–20 connectives prolonged at top, G1; ovule 1, marginal; style single. Nut. *Widespread.*

One genus (*Ceratophyllum*), native to both Europe and North America.

PIPERALES

Flowers inconspicuous, with bracts, often in spikes.

47 **Saururaceae**. Herbs. Leaves alternate, simple, stipulate; ptyxis involute (*Saururus*). Inflorescence spike/raceme/head. Flowers bisexual, actinomorphic. P0, A6–8, G3–5/(3–5) superior/inferior, 1-celled or 3–5-celled, ovules 1–10 per cell, parietal/axile; styles 3–4, free. Follicle/fleshy capsule. *Scattered*.

There are 5 genera and about 7 species. Two genera (*Saururus, Anemopsis*) are native to North America.

48 **Piperaceae**. Herbs/shrubs. Leaves usually alternate, entire, stipulate/exstipulate; ptyxis variable. Inflorescence a fleshy spike. Flowers minute, usually bisexual, often sunk in spike. P0, A2–10, G(2–4), superior; ovule 1, basal; stigma sessile, often brush-like. Small drupe. *Tropics*.

A family of 8 genera and over 3,000 described species. Three genera occur in North America, and species of 2 (*Peperomia, Piper*) are widely cultivated as ornamentals. The fruits of *Piper nigrum* are the source of pepper.

49 **Chloranthaceae**. Herbs/shrubs. Leaves opposite, simple, stipulate; ptyxis conduplicate (*Chloranthus*). Inflorescence spike/panicle/head. Flowers unisexual, actinomorphic; male: A1–3 sometimes made up of a central whole stamen attached to 2 half-stamens; female: P3, epigynous, G1-celled; ovule 1, pendulous; stigma sessile. Drupe. *Tropics, S temperate areas*.

A rather obscure family of 5 genera and about 40 species. One genus is native in the southern parts of North America, and one or two species from two genera are grown as ornamentals.

ARISTOLOCHIALES

Herbaceous or shrubby, often climbing; leaves alternate, exstipulate; ovules axile; capsule; flowers or leaves bizarre.

50 Aristolochiaceae. Herbs/climbers. Leaves alternate, simple, often cordate, exstipulate; ptyxis conduplicate. Inflorescence various. Flowers bisexual, actinomorphic/zygomorphic. PA epigynous. P(3) often bizarre and foetid, A6 often attached to style, G(4–6); ovules n, axile; stigmas several, sessile. Capsule. *Mostly tropics.*

There are 7 genera and about 600 species. Two genera are native to Europe and 3 to North America. Species of *Asarum* and *Aristolochia* are grown for their bizarre flowers.

51 Rafflesiaceae. Root- or branch-parasites lacking chlorophyll. Leaves scale-like/0. Flowers unisexual, actinomorphic. PA epigynous. P4–10/(4–10), A n, G(4–6–8); ovules n, parietal. Berry. *Mostly Old World tropics.*

There are 9 genera and about 55 species. *Cytinus* is native to Europe and *Bdallaphyton* to northern Mexico. None is cultivated. The flowers of *Rafflesia* (Malaysia, Indonesia) are the largest known.

GUTTIFERALES

Usually woody; A mostly n, often united in bundles.

52 Dilleniaceae. Usually woody, sometimes climbing/herbaceous. Leaves alternate, simple, exstipulate; ptyxis conduplicate. Inflorescences various. Flowers bisexual, actinomorphic. KCA hypogynous. K5/4–18 enlarging in fruit, C5, A n sometimes in 2–5 bundles, G1–20 sometimes somewhat coherent; ovules 1–n, marginal; styles free. *Tropics, Australasia.*

A family of about 12 genera and 150 species. One genus is native to North America; species of 2 other genera are occasionally cultivated for ornament.

53 **Paeoniaceae**. Herbs/soft-wooded shrubs. Leaves alternate, compound, exstipulate; ptyxis variable. Flowers usually solitary, bisexual, actinomorphic. KCA hypogynous. K5 persistent and differing in size, C5–9, A n, G2–8; ovules n, marginal; styles free. Large follicles. *N Temperate areas*.

A single genus (*Paeonia*, 20–25 species), native to both Europe and North America, and widely cultivated.

54 **Crossosomataceae**. Shrubs. Leaves alternate, simple, exstipulate. Flowers solitary, terminal, bisexual, actinomorphic. KCA perigynous. K5, C5, A15–n, G3–5; ovules n, marginal; styles free. Follicles; seeds with much-divided arils. *Western N America*.

Another family of a single genus (*Crossosoma*, 3 or 4 species), native to western USA and adjacent Mexico.

55 **Eucryphiaceae**. Trees/shrubs, evergreen in wild, tending to be deciduous in cultivation. Leaves opposite, simple/pinnate, stipulate (stipules falling early); ptyxis revolute. Flowers solitary, bisexual, actinomorphic. KCA hypogynous. K(4) falling as a unit as flower opens, C4, A n, G(5–12) rarely -(18); ovules n, axile; styles free, short. Capsule. *Australia (Tasmania), Chile*.

A further family of a single genus (*Eucryphia*, about 10 species); several species and hybrids are cultivated as extremely ornamental, late-flowering small trees or shrubs.

56 **Actinidiaceae**. Woody, some climbing. Leaves alternate, simple, evergreen/deciduous, exstipulate; ptyxis supervolute or conduplicate. Inflorescences various. Flowers unisexual/bisexual, actinomorphic. KCA hypogynous. K3–8 usually 5, C3–8 usually 5, A n anthers often opening by pores, G(3–5); ovules n, axile; styles usually free, occasionally fused at base. Berry/capsule. *Subtropics & tropics*.

There are 3 genera and about 100 species; species of all 3 are cultivated for ornament. *Actinidia chinensis* is widely grown in warm areas (especially New Zealand) for its edible berries (kiwi fruit).

57 Ochnaceae. Trees/shrubs. Leaves alternate, simple, stipulate, mostly evergreen. Flowers bisexual, actinomorphic. KCA hypogynous. K5, C5–10, A n opening by slits or apical pores, G3–n, united only by a common, single style, which is slightly lobed at the top; ovules 1–n per cell, axile. Schizocarp, often fleshy. *Mostly tropics.*

A family of 37 genera and about 450 species. Two genera are native to North America, and a few species of *Ochna* are cultivated as ornamental greenhouse shrubs.

58 Dipterocarpaceae. Very large trees. Leaves alternate, entire, evergreen, stipulate; ptyxis conduplicate. Inflorescence usually a panicle. Flowers bisexual, actinomorphic. KCA hypogynous. K5, 2 usually enlarging in fruit, C5, A n, G usually (3); ovules usually 2 per cell, axile/rarely parietal. Nut. *Tropical Asia, 1 genus in tropical America.*

There are 16 genera and some 530 species, all of them very large tropical trees; juvenile specimens of a few species are grown in glasshouses in Europe, where they very rarely flower.

59 Theaceae. Trees/shrubs/rarely climbing. Leaves alternate, simple, exstipulate, usually evergreen; ptyxis conduplicate or supervolute. Inflorescence raceme/panicle/flowers solitary. Flowers bisexual, actinomorphic. KCA hypogynous/K hypogynous CA perigynous. K5–6, C5–14, A n filaments sometimes partly united below, united to bases of petals, G(3–6); ovules n, axile; styles 3–6 free/united and single, lobed at the top. Capsule/berry. *Tropics & warm temperate areas.*

A family of about 30 genera and 520 species. Seven genera are native in North America, and species of several others are widely cultivated as ornamental shrubs. The most commonly grown is *Camellia*), and the tea plant, widely grown in subtropical areas, belongs to this genus (*C. sinensis*).

60 Caryocaraceae. Woody. Leaves alternate/opposite, evergreen, of 3 leaflets, stipulate/exstipulate. Inflorescence a raceme. Flowers bisexual, actinomorphic. KCA hypogynous. K(5–6), C5–6/rarely (5–6), A n with long, projecting, coloured filaments, G(4–20); ovules 1 per cell, axile; styles 4–20, free. Drupe/schizocarp. *Tropical America.*

Two genera and 24 species; very uncommon in cultivation.

61 **Marcgraviaceae**. Woody climbers, often epiphytic. Leaves alternate, simple, exstipulate, evergreen; ptyxis supervolute. Inflorescences racemes/umbels; flowers bisexual, zygomorphic, some sterile, their bracts variously modified into pitcher-like, pouched or spurred nectaries. KCA hypogynous. K4–7/(4–7), C4–5/(4–5) falling as a unit when united, A3–n, G(2–n); ovules n, parietal; style single, short or stigmas sessile. Capsule/indehiscent. *Tropical America*.

There are 5 genera and about 110 species. One genus is native to North America, and species of *Marcgravia* are occasionally cultivated as evergreen woody climbers with bizarre inflorescences.

62 **Guttiferae/Clusiaceae**. Herbs/woody, mostly with yellow, red or clear resinous latex. Leaves usually opposite/rarely whorled/alternate, usually exstipulate, gland-dotted; ptyxis flat or conduplicate. Inflorescence cymose. Flowers unisexual/bisexual, actinomorphic. KCA hypogynous. K2–6/(2–6), C2–10, A n often in 3–5 bundles, G(2–10)/1-celled; ovules 1–n per cell, axile/parietal; styles several free, or united below and lobed above, rarely stigmas sessile. Capsule/berry. *Widespread*.

There are 48 genera and over 1,000 species. The family *Hypericaceae*, consisting of herbs or shrubs mainly from temperate areas, with stamens united in 3–5 bundles and separate styles, is often distinguished; it is native to both Europe and North America.

SARRACENIALES

Leaves forming insectivorous pitchers or with insect-trapping and -digesting hairs and glands, or insect-trapping by folding of the 2 halves of the leaf.

63 **Sarraceniaceae**. Herbs. Leaves basal tubular pitchers, exstipulate. Flowers solitary/racemose, bisexual, actinomorphic. KCA/PA hypogynous. K4–6, C5/0, A8–n, G(3–5); ovules n, axile; style 1. Capsule. *America*.

Three genera from swamps of North and South America. Two genera are native to North America, and species of all 3 genera are cultivated.

64 **Nepenthaceae**. Woody climbers/shrubs. Leaves alternate, exstipulate, tips prolonged, terminating in insectivorous pitchers. Inflorescence a raceme/panicle. Flowers unisexual, actinomorphic. PA hypogynous. P3–4, A n, G(3–4); ovules n, axile; stigmas sessile. *Tropical SE Asia.*

A single genus (*Nepenthes*, 70 species). Several species and hybrids are cultivated as pitcher plants.

65 **Droseraceae**. Herbs/small shrubs, rarely aquatic. Leaves in basal rosettes, usually simple, insectivorous with sticky hairs or traps, exstipulate, ptyxis circinate. Flowers bisexual, actinomorphic. KCA hypogynous. K4–5, C4–5, A4–5/10–20 pollen in tetrads, G(3–5); ovules n, parietal; styles 3–5 free/single, lobed at apex. Capsule. *Temperate areas.*

Four genera with well over 100 species (many described recently, especially from Western Australia). Three of the genera are native to Europe and 2 to North America. Species of all 4 genera are cultivated.

PAPAVERALES

Flowers actinomorphic (often with 2 planes of symmetry)/ zygomorphic; ovules parietal; seeds often with arils; milky or coloured latex sometimes present.

66 **Papaveraceae**. Herbs/rarely shrubs, with clear, milky or coloured sap. Leaves alternate/basal/rarely opposite, simple/divided, exstipulate; ptyxis variable. Inflorescence cymose/racemose/flowers solitary. Flowers bisexual, actinomorphic/zygomorphic. KCA hypogynous/rarely perigynous. K usually 2–3/rarely 4–n/rarely united and falling as a whole, often falling early, C0–n, usually 4/6, occasionally variously united above the base, A n/4–6 sometimes united, G(2–n) rarely almost free; ovules 1–n, parietal; style 1 or styles very short, stigmas several, sessile or almost so. Capsule. *Mostly N temperate areas.*

There are 41 genera with over 600 species, with many genera native to either or both Europe and North America. It is often divided into two segregate families (rarely into three, the third, *Hypecoaceae*, included here in *Fumariaceae*).

1a Sap milky or coloured; stamens usually numerous; nectaries absent
66a **Papaveraceae**

b Sap not milky or coloured; stamens usually 4, often arranged in 2 groups of 2 half-stamens united with and on either side of a whole stamen; nectary present   66b **Fumariaceae**

66a **Papaveraceae** in the strict sense. Mostly herbs. Sap usually milky or coloured. Flowers actinomorphic. KCA hypogynous/perigynous, flowers without nectar. C-segments never spurred, A usually n, rarely as few as 6, when free, G(2–n), styles very short, stigmas several, sessile or almost so. *N temperate areas & S America.*

There are some 23 genera; native to both Europe and North America. Many species are cultivated, especially in the genera *Meconopsis* and *Papaver*. The family is economically very important because of the opium poppy (*Papaver somniferum*), from which important drugs are extracted legally (morphine, codeine), or illegally (heroin).

66b **Fumariaceae**. Always herbs, occasionally climbing; sap clear. Flowers zygomorphic/with 2 planes of symmetry. At least 1 C-segment spurred and with a nectary projecting backwards into the spur; A4, made up of 2 groups of 2 half-stamens united to a whole stamen; G usually (2), style well-developed, sometimes elaborate. *N temperate areas.*

Eighteen genera; 8 (1 introduced) occur in Europe, and about 5 in North America. Many species of *Corydalis* and *Dicentra* are cultivated as ornamentals.

67 **Capparaceae**. Woody/herbaceous. Leaves alternate, simple/compound, stipulate/exstipulate; ptyxis conduplicate. Flowers solitary/racemes, somewhat zygomorphic, bisexual. KCA hypogynous. K4–8 usually 4, C4–8 usually 4, A4/6–n, G(2) often stalked; ovules few–n, parietal; style short, stigma lobed/styles sessile. Capsule/berry/nut. *Tropics, warm temperate areas.*

A family of 45 genera and about 700 species. Two genera are native to Europe and 9 to North America. Species of 1 or 2 genera are cultivated as ornamentals. The fruits of *Capparis spinosa* provide the capers of commerce.

68 **Cruciferae/Brassicaceae**. Usually herbs. Leaves usually alternate, simple/divided, exstipulate/rarely with basal lobes appearing like stipules; ptyxis variable. Inflorescence usually without bracts, spike/raceme/flowers solitary. KCA hypogynous/rarely perigynous. K4, C4/rarely 0, A usually 6, 2 shorter and 4 longer/rarely fewer or more, G(2); ovules 2–n, parietal; stigmas sessile or style single, short. Usually capsule with false septum/indehiscent. *Cosmopolitan in cooler areas.*

A large and rather uniform family of 390 genera and 3,000 species. One hundred and eight genera are native to Europe, and 94 to North America. Many are grown as ornamentals, others as vegetable crops (species of *Brassica*), salads (*Eruca*, *Lepidium*, *Nasturtium*) or as spices (*Sinapis*).

69 **Resedaceae**. Herbs/shrubs. Leaves alternate, simple/divided, stipules minute. Inflorescence a spike/raceme. Flowers unisexual/bisexual, zygomorphic. KCA usually hypogynous. K4–8, C2–8 usually fringed, A3–n, G(2–6)/rarely 2–6, open at top; ovules n, parietal; styles absent, stigmas sessile. Capsule/follicle/berry, usually open and gaping at apex. *N temperate areas.*

Six genera with 75 species. Two genera are native to Europe, 2 to North America. A few species of *Reseda* are grown as ornamentals or for their scented inflorescences.

70 **Moringaceae**. Trees. Leaves alternate, 2–3-pinnate, exstipulate though with glands at base of stalk. Flowers bisexual, zygomorphic in panicles. KCA perigynous. K5, C4–5, A5 with 3–5 staminodes, anthers 1-celled, G(3); ovules n, parietal; style 1 long, slender. Capsule, 3-sided. *Mostly Old World tropics.*

A family of a single genus (*Moringa*, 12 species), occurring mainly in desert areas. A few species are grown as ornamentals.

BATALES

Dioecious shrubs; flowers in catkins; fruit a syncarp of berries.

71 **Bataceae**. Shrubs. Leaves opposite, simple, exstipulate. Inflorescence a catkin. Flowers unisexual; male P2, A4–5 with 4–5 staminodes; female: P0, G(4) naked; ovules 1 per cell, ascending; styles free. Syncarp of berries. *Coasts of America.*

A family of a single genus (*Batis*) with 2 species occurring in coastal regions of both North and South America.

ROSALES

Woody/herbaceous; leaves often compound; carpels free/united.

72 **Platanaceae**. Trees with exfoliating bark, often with stellate hairs. Leaves alternate, lobed, stipulate, leaf-bases covering axillary buds; ptyxis conduplicate-plicate. Inflorescence a raceme of hanging spherical heads. Flowers unisexual, actinomorphic. PA perigynous/hypogynous. P3–5/(3–5) sometimes considered as bracts, A3–4, G5–9; ovules 1/rarely 2, marginal; styles free. Fruit prickly balls of achenes with persistent styles. *North temperate areas.*

A single genus (*Platanus*) with about 8 species, occurring in both Europe and North America. Several species and hybrids are cultivated, especially as street trees.

73 **Hamamelidaceae**. Woody, often with stellate hairs. Leaves usually alternate, simple/lobed, stipulate; ptyxis flat or conduplicate, rarely supervolute. Flowers in spikes, clusters or pairs, unisexual/bisexual, actinomorphic/zygomorphic. KCA perigynous/epigynous. K4–5/(4–5), C4–5/rarely fewer or more, A4–5/rarely more, anthers opening by valves, G(2); ovules 1–n per cell, axile; styles 2, free. Woody capsule. *Tropics & subtropics, mainly E Asia, few temperate.*

A family of 28 genera with 90 species. Three genera are native to North America. Species of several genera are widely cultivated as flowering (many very early-flowering) shrubs, especially in the genera *Corylopsis* and *Hamamelis*.

74 **Crassulaceae**. Herbs/shrubs, with succulent leaves. Leaves alternate/opposite, simple, exstipulate. Flowers in cymes/spikes/panicles/racemes/rarely solitary, bisexual, actinomorphic. KCA hypogynous. K4–7/12–16/(4–5), C4–7/12–16/(4–5), A4–5/10–12/rarely more, G4–5/10–12; ovules n, marginal; styles free. Follicles. *Widespread (except Australia)*.

A family of 30 genera and about 1,400 species, all succulent; almost all of the genera are in cultivation. Thirteen genera are native to Europe and 12 to North America.

75 **Cephalotaceae**. Herbs. Leaves alternate in a rosette, modified into stalked, insectivorous pitchers, exstipulate. Flowers in racemes, bisexual, actinomorphic. PA perigynous. P6, A12, connective swollen, glandular, G6; ovules 1 per carpel, basal/marginal; styles free. Follicles. *Australia*.

A single genus (*Cephalotus*) with a single species, *C. follicularis*, which is occasionally cultivated in Europe.

76 **Saxifragaceae**. Herbs. Leaves alternate, all basal, usually simple, exstipulate/rarely with small stipules; ptyxis variable. Flowers solitary/racemes, usually bisexual, actinomorphic/rarely zygomorphic. KCA perigynous/epigynous. K4–5, C4–5/rarely 0, A4–5/8/10, G(2)/more occasionally almost free; ovules n, axile; styles 2/rarely more, free, usually divergent/occasionally stigmas sessile. Capsule. *Temperate areas*.

In this broad sense, a family of 80 genera and about 1,200 species. Genera are native to both Europe and North America, and many are cultivated for ornament.

The broad family is now usually divided into several segregate families; different authorities recognise different selections of these (there are several more than those included below).

1a	Plants woody	2
b	Plants herbaceous	5
2a	Leaves of 3 leaflets, stalkless, opposite, evergreen	76d **Baueraceae**
b	Leaves not as above	3
3a	Stamens 8 or more; leaves usually opposite	76e **Hydrangeaceae**

b	Stamens 4–6; leaves alternate	4

4a Disc present; leaves usually with gland-tipped teeth

76f **Escalloniaceae**

b Disc absent; leaves without gland-tipped teeth

76g **Grossulariaceae**

5a Stamens not alternating with staminodes   76a **Saxifragaceae**

b Stamens alternating with staminodes   6

6a Ovary spherical; stamens 5; petals 5; staminodes much divided, gland-tipped   76b **Parnassiaceae**

b Ovary 4-sided, cylindric; stamens 4 or 8, petals 4; staminodes simple   76c **Francoaceae**

76a **Saxifragaceae**. Herbs. Leaves alternate/basal, exstipulate, often evergreen and forming rosettes. Flowers in racemes, bisexual, usually actinomorphic/rarely zygomorphic. KCA perigynous (sometimes very slightly so)/epigynous. K usually 4–5, C usually 4–5, A usually 8/10/less often 4–5, G(2)/rarely 2/(4 or more); ovules n, axile/parietal; styles 2, free. Capsule. *Mainly N temperate areas*.

About 30 genera and 475 species, with genera native to both Europe and North America. Many of the genera are widely cultivated, especially species and hybrids of the largest genus, *Saxifraga*.

76b **Parnassiaceae**. Herbs. Leaves mostly basal, simple, exstipulate. Flowers solitary, actinomorphic, bisexual. KCA hypogynous. K5/(5), C5, A5 alternating with 5 staminodes, G(3–4); ovules n, parietal; style absent, style 1, short/stigmas sessile. Capsule. *Arctic & N temperate areas*.

A single genus (*Parnassia*), with species native to both Europe and North America; it is difficult in cultivation and rarely seen.

76c **Francoaceae**. Herbs. Leaves mostly basal, simple/divided, exstipulate, more or less evergreen. Racemes. Flowers bisexual, actinomorphic. KCA hypogynous. K(4), C4 equal or unequal, A8, G(4); ovule n, parietal; style 1, stigmas lobed. 4-sided capsule. *Chile*.

Two genera, of which one (*Francoa*) is commonly cultivated as an ornamental.

76d **Baueraceae**. Shrubs. Leaves opposite, of 3 leaflets, exstipulate, evergreen. Flowers solitary, bisexual, actinomorphic. KCA slightly perigynous. K4–10, C4–10, A4–10, anthers with pore-like slits, G(2); ovules n, axile; styles 2, free. Capsule. *Australia*.

Another small family of a single genus (*Bauera*), which is occasionally cultivated as an ornamental greenhouse shrub.

76e **Hydrangeaceae**. Herbs/softly wooded shrubs, rarely climbing, many with stellate hairs. Leaves usually opposite, simple, exstipulate. Inflorescences various. Flowers mostly bisexual (sometimes outer flowers of inflorescence sterile), actinomorphic. KCA hypogynous/perigynous/epigynous. K4–5, C4–7, A4–n, G(2–7)/rarely 1-celled; ovules n, axile/parietal; styles 1, stigma head-like/2–7, free or almost so. Capsule/berry. *Widespread*.

As treated here, this family includes the *Philadelphaceae* (shrubs with stellate hairs; filaments often toothed beside the anther), which is sometimes treated as distinct. There are about 17 genera and 170 species, several of the genera native to either Europe or North America. Many of the genera are cultivated as ornamental shrubs, especially *Deutzia*, *Hydrangea* and *Philadelphus*.

76f **Escalloniaceae**. Trees/shrubs. Leaves mostly alternate, evergreen, exstipulate, with gland-tipped teeth. Flowers in racemes, actinomorphic, bisexual. KCA perigynous/epigynous. K(4–6) usually (5), C4–6 usually 5, A4–6 usually 5, G usually (2); ovules n, parietal; style 1, somewhat lobed at apex. Capsule/berry. *Mainly S hemisphere*.

There are about 15 genera, with about 70 species. Some species and hybrids of the largest genus, *Escallonia* are cultivated as ornamental shrubs and as hedging plants. This family is sometimes further divided into *Iteaceae*, *Brexiaceae*, etc.

76g **Grossulariaceae**. Shrubs, often spiny. Leaves alternate, simple/lobed, stipulate/exstipulate, usually deciduous. Flowers in racemes, unisexual/bisexual, actinomorphic. KCA epigynous. K4–5, C4–5, A4–5, G(2); ovules few–n, parietal; styles free/united into a single style lobed at apex. Berry. *Temperate N hemisphere, S America*.

[121]

A single genus (*Ribes*), with about 150 species, including species native to both Europe and North America. Many species are cultivated as ornamental shrubs, and a few are cultivated for their edible fruit (blackcurrant, redcurrant).

77 **Cunoniaceae**. Woody. Leaves evergreen, opposite/whorled, usually pinnate, stipulate; ptyxis conduplicate or supervolute. Flowers in racemes/heads, unisexual/bisexual, actinomorphic. KCA perigynous/epigynous. K4–5/rarely (4–5), C4–5/rarely 0, A8/10, G(2–3); ovules n, axile; styles 2–3. Capsule/rarely drupe/nut. *Mostly S hemisphere.*

There are 24 genera with about 350 species. Species of 7 genera are occasionally cultivated as glasshouse shrubs.

78 **Pittosporaceae**. Woody, sometimes climbing. Leaves alternate/opposite, simple, exstipulate, often evergreen; ptyxis variable. Inflorescence clusters/flowers solitary. Flowers bisexual, actinomorphic. KCA hypogynous. K5, C5, A5 anthers sometimes united, G(2–5); ovules n, axile; style 1, short, stigmas 2–5. Capsule/berry. *Tropics & S temperate Old World.*

Nine genera and 240 species. Species of about 4 genera are cultivated as ornamental shrubs, especially *Pittosporum tobira.*

79 **Byblidaceae**. Herbs with insect-trapping glandular hairs on leaves. Leaves alternate, linear, spirally coiled when young, exstipulate. Flowers solitary, bisexual, actinomorphic. KCA hypogynous. K5, C5, A5 anthers opening by pores, G(2); ovules n, axile. Capsule. *N & W Australia.*

A single genus (*Byblis*) with 2 species, which are increasingly being cultivated by insectivorous plant enthusiasts.

80 **Roridulaceae**. Small shrubs. Leaves with insect-trapping, glandular hairs, alternate but clustered at ends of branches, exstipulate, entire/lobed. Flowers solitary/in racemes, bisexual, actinomorphic. KCA hypogynous. K5, C5, A5 anthers opening by pores, G(3); ovules 1–4 per cell, axile; style 1, stigma capitate. Capsule. *South Africa.*

Another monogeneric family (*Roridula*), sometimes included in the *Byblidaceae*. Again, becoming increasingly cultivated.

81 Bruniaceae. Low heather-like shrublets. Leaves alternate, needle-like, exstipulate. Flowers in spikes/heads, bisexual, actinomorphic. KCA hypogynous/epigynous. K4–5/(4–5), C4–5, A4–5, G(2–3); ovules 1–2 per cell; styles 2–3, joined at extreme base. *South Africa*.

A family of 11 genera and 69 species. A few species of *Brunia* have, at times, been cultivated for the cut-flower trade in Europe.

82 Rosaceae. Herbs/shrubs/trees/rarely climbing. Leaves usually alternate, rarely opposite/whorled, usually stipulate, simple/divided, evergreen/deciduous; ptyxis very variable (important in the classification of the species of *Prunus*). Inflorescences various. Flowers usually actinomorphic and bisexual. KCA perigynous (perigynous zone sometimes very small)/epigynous. K4–6/(4–6) epicalyx sometimes present, C4–6/rarely 0, A4–n, G1–n/(2–5); ovules 1–many, axile/marginal; styles as many as carpels, free or almost so/united for more than half their length. Fruit variable, follicles/achenes/'berry' of drupelets/pome. *Cosmopolitan*.

A large and very variable family of 115 genera and 3,200 species. Many genera are native to Europe and/or North America, and very many are cultivated as ornamental herbs, shrubs or trees (e.g. *Cotoneaster*, *Crataegus*, *Potentilla*, *Prunus*, *Rosa*, *Rubus*, *Spiraea*, etc. Many produce edible fruit, notably *Malus* (apple), *Pyrus* (pear) and *Prunus* (cherry).

83 Chrysobalanaceae. Woody. Leaves alternate, stipulate, often leathery. Flowers in racemes/panicles/cymes, usually bisexual, actinomorphic/zygomorphic. KCA perigynous. K5, C4–5, A3–n, G(2–3) often 1 or 2 carpels sterile, often asymmetrically placed in the perigynous zone; ovules 2, basal; styles free. Drupe. *Mainly tropics*.

There are 500 species in 17 genera. A few species of *Chrysobalanus*, *Maranthes* and *Parinari* are occasionally cultivated in glasshouses.

84 Leguminosae/Fabaceae. Herbs/woody/climbing. Leaves usually alternate, stipulate, pinnately compound, sometimes 2–3-pinnate, or of a single or 2–3 leaflets; ptyxis of leaflets almost always conduplicate, but supervolute in some species of *Lathyrus*. Inflorescence various, usually a raceme. Flowers usually bisexual, actinomorphic/zygomorphic. KCA hypogynous/perigynous. K usually 5/(5), C4–5 sometimes some of them united at or above the base/(4–5)/rarely 0; A4–n, usually 10, free or variously united, anthers sometimes opening by pores, G1/rarely 2–15; ovules 1–n, marginal; styles as many as carpels. Legume (sometimes indehiscent)/lomentum/nut. *Cosmopolitan.*

A very large and relatively uniform family of about 600 genera and 13,500 species. Many genera are native to Europe and/or North America, and many are cultivated. It is often divided into 3 separate families (which are considered as subfamilies when the Leguminosae is retained as the family).

1a Corolla actinomorphic; petals valvate in bud; stamens 4–many; leaves bipinnnate, rarely reduced to phyllodes; seeds with U-shaped lateral line (*mainly tropics & subtropics*)

84a Mimosaceae

 b Corolla zygomorphic (sometimes weakly so); petals imbricate in bud, rarely absent; stamens 10 or fewer; leaves simply pinnate, of 3 leaflets or simple; seeds usually without a lateral line, rarely with a closed lateral line   2

2a. Upper petal interior, or petal 1 or petals absent; seed usually with a straight radicle (*mainly tropical*)   **84b Caesalpiniaceae**

 b Upper petal exterior; seed usually with an incurved radicle (*widespread*)   **84c Papilionaceae/Fabaceae**

84a Mimosaceae. Shrubs or trees, more rarely herbaceous. Leaves usually bipinnate, rarely tripinnate or reduced to expanded stalks and rachises (phyllodes). C actinomorphic, segments free or more commonly fused, valvate in bud; A4–many, anthers opening by slits, pollen sometimes not granular. Seeds with U-shaped lateral line. *Mainly tropics & subtropics.*

Fifty-eight genera, 3,100 species. Many are grown as ornamentals in Europe, especially species of *Acacia*, *Mimosa* and *Albizia*.

84b **Caesalpiniaceae.** Trees, shrubs or herbs. Leaves usually pinnate, rarely bipinnate, never reduced to phyllodes. C zygomorphic, petals free, imbricate in bud such that the uppermost petal is inside all the others; A3–10, often deflexed downwards, anthers often opening by pores, pollen granular. Seeds usually without a lateral line. *Mainly tropics.*

There are 162 genera and about 2,000 species. Species of a few genera are native to Europe and/or North America, and species of *Bauhinia, Cercis* and *Gleditsia* are commonly grown as glasshouse or hardy shrubs or trees. In the tropics, species of *Delonix* and *Amherstia* form spectacular ornamentals and these are occasionally seen in glasshouses in Europe.

84c **Papilionaceae/Fabaceae.** Herbs, shrubs or trees. Leaves simple to trifoliolate to pinnate or palmate, rarely completely absent or very reduced. C zygomorphic, petals free or some of them united, imbricate in bud such that the uppermost is outermost and often larger than the rest, the 2 lower close or united and forming a keel; A usually 10, filaments free or all united, or that of the uppermost free, all the others united into a sheath; anthers opening by slits. Seeds usually without a lateral line. *Cosmopolitan.*

There are about 450 genera and 11,300 species. Many genera are native to Europe and/or North America, and many are grown as ornamentals. Many of the genera produce crops of economic importance.

85 **Krameriaceae.** Shrubs/herbs. Leaves alternate, entire, exstipulate. Inflorescence axillary/racemose. Flowers bisexual, zygomorphic. KCA hypogynous/K hypogynous CA perigynous. K4–5, C5/(5) the lower pair often modified into glands, A3–4 anthers opening by pores, G1-celled; ovules 2, pendulous; style 1, stigma disc-like. Fruit 1-seeded, indehiscent, covered with barbed spines. *Mainly tropical America.*

A single genus (*Krameria*) with about 15 species, which occurs in the southern part of North America. One of the species is occasionally seen in cultivation in Europe.

PODOSTEMALES

Much-modified aquatics.

86 Podostemaceae. Aquatics of running water, often resembling algae, mosses or hepatics. Leaves alternate/rarely 0, simple. Flowers bisexual, zygomorphic. PA hypogynous. P2–3/(2–3), A1–4, G(2); ovules n, axile; styles 2, free, short. Capsule. *Mainly tropics*.

There are 50 genera with 275 species of very bizarre plants. They are mainly tropical, but 1 genus is found in North America.

GERANIALES

Mostly herbs; leaves simple/divided; G superior; dehiscence of fruit often explosive.

87 Limnanthaceae. Herbs. Leaves alternate/basal, divided, exstipulate; ptyxis conduplicate. Flowers solitary, bisexual, actinomorphic. KCA hypogynous. K3–5, C3–5, A6–10, G(3–5), bodies of carpels free, style 1, divided above into as many stigmas as there are carpels; ovules 1 per cell, ascending. Nutlets. *Temperate N America*.

Two genera with 6 species, native to North America (California). Species of *Limnanthes* are widely cultivated as half-hardy annuals.

88 Oxalidaceae. Herbs/shrubs/trees. Leaves alternate/basal, pinnate, palmate or with 3 leaflets, exstipulate. Inflorescences various. Flowers bisexual, actinomorphic. KCA hypogynous. K5/(5), C5 contorted, A10 filaments sometimes joined into a tube, G(5); ovules 1– more, axile; styles 5, free. Capsule, often explosive. *Widespread*.

A family of 8 genera with 575 species. One genus (*Oxalis*) is native to both Europe and North America. Many species of this genus are also in cultivation.

89 Geraniaceae. Usually herbs. Leaves alternate/opposite, simple/ compound, stipulate/exstipulate; ptyxis conduplicate-plicate. Inflorescences various. Flowers bisexual/rarely unisexual, actinomorphic/ zygomorphic. KCA usually hypogynous. K3–5/(3–5) sometimes spurred with spur attached to flower-stalk, C2–5, A(5–15) often some infertile, G(3–5) often long-beaked; ovules 1–n per cell, axile; style 1, divided at the top/3–5, free. Capsule/berry/schizocarp. *Widespread*.

There are 14 genera with over 700 species. Species of *Geranium* are native to Europe and North America, and species of *Erodium* to Europe. Many species are cultivated, especially from the genera *Erodium*, *Geranium* and *Pelargonium*.

90 **Tropaeolaceae**. Herbs. Leaves alternate/opposite, simple/divided, stipulate/exstipulate; ptyxis flat or conduplicate. Flowers bisexual, zygomorphic, solitary, axillary. KC partly perigynous, A hypogynous. K5 spurred, C2/5, A8, G(3); ovule 1 per cell, axile; style 1, lobed at the apex. Schizocarp. *C & S America*.

A single genus (*Tropaeolum*) with about 80 species. Several of the species are cultivated as ornamental climbers or scramblers.

91 **Zygophyllaceae**. Herbs/shrubs. Leaves usually opposite, usually compound, stipulate, often fleshy; ptyxis variable. Inflorescence cymose/flowers solitary. Flowers usually bisexual and actinomorphic. KCA hypogynous, disc usually present. K4–5, C4–5, A5–15, G(2–5); ovules n, axile; style 1, stigma terminal/stigma sessile. Capsule/drupe-like/schizocarp. *Tropics & warm temperate areas*.

A rather variable family, with 27 genera and 250 species, mostly desert plants, sometimes succulent. Six genera are native to Europe, 7 to North America. Only a small number is cultivated.

92 **Linaceae**. Herbs/shrubs. Leaves alternate/opposite, entire, stipulate/exstipulate; ptyxis flat or conduplicate. Inflorescence a cyme. Flowers bisexual, actinomorphic. KCA hypogynous. K4–5/(4–5) C3–5, A4–5/10/15 sometimes united at base, G(3–5) often 6–10-celled with 3–5 secondary septa; ovules 1–2 per cell, axile; styles 3–5, free. Capsule/drupe. *Widespread*.

A mainly tropical family of 15 genera and over 500 species. Two genera are native to Europe and 4 to North America. A few genera are cultivated for their flowers (*Linum*, *Reinwardtia*), and flax and linseed oil are extracted from *Linum usitatissimum*.

93 **Erythroxylaceae**. Woody. Leaves alternate, simple, stipulate; ptyxis revolute. Inflorescences various. Flowers bisexual, actinomorphic. KCA hypogynous. K5, C5 each with an appendage on inner face, A(10) filaments united at base, G(3) often only 1 cell developing; ovules 1–2 per cell, axile; styles free. Fruit berry-like. *Tropics (mostly America)*.

A small, rather uniform family of 4 genera and 250 species. One genus is native in North America. *Erythroxylon coca*, the original source of the drug cocaine, is sometimes grown as a curiosity in European glasshouses.

94 **Euphorbiaceae**. Woody/herbs/succulents, milky sap often present. Leaves usually alternate and stipulate, simple/compound, rarely absent, sometimes replaced by cladodes; ptyxis very variable. Inflorescences various, sometimes within a cup with glandular margins (cyathium). Flowers unisexual, actinomorphic. PA/KCA hypogynous. P4–6/(2–6)/rarely K5/10, C5/(5), A1–n/(2–n), G(2–4)/ rarely more, usually (3); ovules 1–2 per cell, axile; styles 2–4/rarely more, usually 3, free or slightly joined at base, often divided above. Fruit usually schizocarpic, seeds often carunculate. *Widespread*.

A very large and highly variable family of 326 genera and 7,750 species. Seven genera are native to Europe, 47 to North America. Species of about 20 genera are cultivated for ornament; the most important is *Euphorbia*, which contains both normal herbaceous and shrubby species and succulent shrubs.

95 **Daphniphyllaceae**. Trees/shrubs. Leaves alternate, crowded, entire, exstipulate, usually evergreen; ptyxis flat. Flowers in axillary racemes, unisexual, actinomorphic. PA hypogynous. Male: P3–8, imbricate, A6–12; female: P0, staminodes few, small/0, G(2) imperfectly divided, styles 1–2, persistent, undivided; ovules 2 per cell, pendulous. Drupe, 1-seeded. *Temperate E Asia*.

A small family with a single genus (*Daphniphyllum*) and about 10 species. One or 2 species of *Daphniphyllum* are cultivated as small evergreen trees.

Mostly woody, often aromatic; stamens usually twice as many as petals.

96 **Rutaceae**. Woody/herbaceous. Leaves alternate/opposite, simple/compound, exstipulate, usually aromatic, gland-dotted, often evergreen; ptyxis usually conduplicate, rarely flat. Inflorescences various. Flowers usually bisexual, usually actinomorphic. KCA usually hypogynous, disc usually present. K3–6/(3–6), C3–6/(3–6)/rarely 0, A3–12/rarely more, staminodes sometimes present, G(4–5)/rarely (n), carpels often free but united by single style/styles rarely free, as many as carpels, ovary often borne on a short stalk; ovules 1–n, axile. Fruit fleshy/capsule/samaras. *Tropics, warm temperate areas.*

A highly variable family of 160 genera and 1,650 species, with 6 genera occurring in Europe (3 introduced) and 20 native to North America. Most species in the family have characteristic glands in the leaves which contain aromatic oils giving rise to identifiable scents when crushed. Species of many genera are cultivated for ornament, and the citrus fruits (oranges, lemons, grapefruit, limes, etc.) belong to the genus *Citrus*.

97 **Cneoraceae**. Shrubs, sometimes with medifixed hairs. Leaves alternate, simple, exstipulate; ptyxis conduplicate. Inflorescence a cyme. Flowers bisexual, actinomorphic. KCA hypogynous, disc 0. K3–4, C3–4, A3–4, G(3–4) on a stalk; ovules 1–2 per cell, axile; styles 3–4 free. Schizocarp. *Mediterranean area, Canary Islands, Cuba.*

A small family of 1 genus (*Cneorum*, sometimes divided into 2 genera) with 3 species. The distribution of the family is unusual and striking. The 1 genus is native to Europe, and *C. tricoccum* is occasionally cultivated as a small flowering shrub requiring glasshouse protection in northern Europe.

98 **Simaroubaceae**. Woody. Leaves alternate, simple/compound, exstipulate; ptyxis conduplicate. Inflorescences various. Flowers usu-

ally unisexual, actinomorphic. KCA hypogynous, disc present. K(3–8), C0–8, A4–14/rarely more, G(2–5); ovules 2 per cell, axile; styles 1–5, free. Fruits various, often samaras. *Mainly tropics*.

A diverse family of 23 genera and about 170 species, 9 genera native to North America, 1 introduced in Europe. A few are cultivated as ornamental trees, notably *Ailanthus altissima*.

99 **Burseraceae**. Woody with aromatic resins. Leaves alternate, usually compound, exstipulate. Inflorescences panicles/flowers solitary. Flowers unisexual/bisexual, actinomorphic. KCA hypogynous, with disc. K(3–5), C3–5/rarely 0, A6–10, G(2–5); ovules 2 per cell, axile; style 1 or stigma sessile. Drupe/capsule. *Mainly tropics*.

There are 18 genera with 540 species; 3 genera are native to North America. A few species of *Boswellia* and *Bursera* are cultivated as aromatic shrubs.

100 **Meliaceae**. Trees, wood often scented. Leaves usually alternate, mostly pinnate, exstipulate; ptyxis conduplicate. Inflorescences cymose panicles. Flowers usually bisexual, actinomorphic/rarely slightly zygomorphic. KCA hypogynous with disc. K(4–6), C4–6/rarely -8, A(8–12)/rarely free/fewer/more numerous, G(2–20); ovules usually 2 or more per cell, axile; style 1, short, stigma terminal. Berry/capsule/drupe. *Mainly tropics*.

A family of 51 genera and about 575 species. Five genera occur in North America, and 1 is introduced in Europe. Species of about 7 genera are grown in glasshouses as ornamental shrubs. and *Melia azedirachita* is grown as a street tree in southern Europe (and elsewhere).

101 **Malpighiaceae**. Woody, often with medifixed hairs. Leaves usually opposite/rarely alternate/whorled, simple/rarely lobed/divided, stipulate/exstipulate, usually evergreen; ptyxis flat or conduplicate. Inflorescences racemes/umbels/panicles. Flowers usually bisexual, slightly zygomorphic/rarely actinomorphic. KCA hypogynous. K5 often some or all sepals with 2 glands on the back, C5 often fringed and/or petals unequal, A(10) anthers sometimes opening by pores, sometimes some sterile, G(3); ovules 1 per cell, axile; styles 3 free,

or all united. Fruits various, often winged mericarps. *Mainly tropical America.*

Sixty-six genera and 1,100 species, with 11 genera occurring in North America. Species of about 10 genera are cultivated, mainly as flowering glasshouse shrubs.

102 **Tremandraceae.** Shrublets. Leaves usually opposite, entire, exstipulate. Flowers solitary, bisexual, actinomorphic. KCA hypogynous. K4–5, C4–5, A8–10 anthers opening by pores, G(2); ovules 1–2/rarely 3 per cell, axile; style 1, short, stigma terminal. Capsule. *Australia.*

There are 3 genera with 46 species. A few species have been cultivated in Europe, but they are rarely seen.

103 **Polygalaceae.** Herbs/shrubs. Leaves usually alternate, entire, exstipulate; ptyxis flat or supervolute. Inflorescences racemes/spikes/panicles. Flowers bisexual, zygomorphic. KCA hypogynous/K hypogynous CA perigynous. K usually 5, lateral pair petal-like, C usually 3 often joined to staminal tube, A(8–10)/rarely (4–5) anthers opening by pores, G usually (2); ovules usually 1 per cell, axile; style 1, stigmas as many as carpels. Fruit usually a capsule; seeds with arils. *Widespread.*

A mainly tropical family of 18 genera and 950 species. One genus is native to Europe and 3 to North America. Species of *Polygala* are found in cultivation.

SAPINDALES

Woody; A rarely on the same radii as the petals; disc often present between perianth and ovary.

104 **Coriariaceae.** Shrubs, branches angular. Leaves opposite, entire, exstipulate; ptyxis flat. Flowers solitary or in racemes, usually bisexual, actinomorphic. KCA hypogynous. K5, C5 keeled inside, A10, G5–10; ovules 1 per carpel, apical; styles as many as carpels. Fruit an achene surrounded by fleshy C. *Scattered.*

[131]

A small family of 1 genus with 5 species. *Coriaria* is native to Europe, and 2 or 3 of its species are found in cultivation.

105 **Anacardiaceae.** Woody with resinous bark. Leaves alternate, simple/compound, exstipulate; ptyxis conduplicate or rarely flat. Inflorescences racemes/panicles. Flowers bisexual/unisexual, actinomorphic. KCA hypogynous, disc often present. K3–5/(5), C3–5 rarely 0, A1–10, G(3–5); ovule 1 per cell, apical/basal; style always 1, sometimes divided above. Drupe, 1-seeded. *Tropics & warm temperate areas.*

A mainly tropical family of resinous trees and shrubs with 73 genera and 850 species. Four genera (1 introduced) occur in Europe and 11 in North America. Species of about 10 genera are cultivated in glasshouses. *Mangifera indica* produces the mango, and *Anacardium occidentale* the cashew nut.

106 **Aceraceae.** Woody. Leaves usually opposite, simple/compound, exstipulate; ptyxis conduplicate-plicate. Flowers clustered/in racemes, mostly unisexual (superficially bisexual), actinomorphic. KC perigynous, A hypogynous/perigynous, disc present. K4–5, C4–5/rarely 0, A4–5, G(2–3); ovules 2 per cell, axile; styles joined in their lower half. Winged mericarps. *N temperate area.*

There are 2 genera, *Acer* with over 100 species, *Dipteronia* with 2; *Acer* is native to both Europe and North America, while *Dipteronia* is from western China. Species of both genera are widely grown as ornamental trees.

107 **Sapindaceae.** Trees/shrubs/woody climbers with tendrils in inflorescence. Leaves usually alternate and compound, stipulate/exstipulate; ptyxis conduplicate. Inflorescences cymes/racemes/panicles. Flowers unisexual (often appearing bisexual), actinomorphic/zygomorphic. KCA hypogynous, disc outside A, often elaborate, K4–5 often unequal, C4–5/rarely 0, A5–8, G usually (3); ovules 2 per cell, axile; style 1, short, sometimes lobed above. Fruits various, seeds with arils. *Mostly tropics.*

A mainly tropical family of 145 genera and 1,300 species. One genus is introduced into Europe, and 17 are native to North

America. Species of 8 genera are cultivated, mainly as glasshouse climbers or trees, though species of *Koelreuteria* and *Xanthoceras* are hardy flowering trees in much of Europe.

108 Hippocastanaceae. Woody. Leaves opposite, palmate, exstipulate; ptyxis conduplicate. Inflorescence racemose. Flowers usually bisexual, zygomorphic. KC perigynous, A usually hypogynous, disc present. K(4–5), C4–5, A5–9, G(3); ovules 2 per cell, axile; style 1. Capsule, seeds large. *Tropics & temperate areas.*

A uniform family of 2 genera with about 15 species. The larger genus, *Aesculus* is native in both Europe and North America, and several of its species are cultivated.

109 Sabiaceae. Woody. Leaves alternate, simple/compound, evergreen/deciduous, exstipulate; ptyxis conduplicate. Inflorescences panicles. Flowers bisexual, zygomorphic, small. KCA hypogynous, disc small. K3–5/(3–5), C5, 3 larger, 2 smaller, A usually 2–5 on the same radii as petals, often only 2 fertile, G(2); ovules 2 per cell, axile; style 1, short, stigma simple. Berry/drupe. *Mostly tropics.*

There are 4 genera with about 50 species. One genus is native to North America, and several species of *Sabia* are cultivated. The family is easily recognised by the fact that the stamens are fewer in number than the petals, but the stamens that are present are on the same radii as some of the petals.

110 Melianthaceae. Herbaceous/woody. Leaves alternate, pinnate, stipulate, stipules between leaf-stalk and stem, often large; ptyxis conduplicate or conduplicate-plicate. Inflorescences racemes. Flowers unisexual/bisexual, zygomorphic. KC perigynous, A hypogynous, with disc. K4–5, C4–5, A4–5/rarely 10, free/united, G(4–5); ovules 1–n per cell, axile; style 1, stigma 4–5-lobed. Capsule. *Africa.*

Three genera with 16 species. Species of *Greyia* and *Melianthus* are cultivated, those of the former requiring glasshouse protection in northern Europe.

111 Balsaminaceae. Herbs. Leaves alternate/whorled, simple, exstipulate; ptyxis involute. Flowers solitary/clusters/racemes (occasionally umbel-like). Flowers bisexual, zygomorphic. KCA hypogynous. K3/5, often coloured, the lowest spurred; C5 (the laterals sometimes fused in 2 pairs), A(5), G(5); ovules n, axile; style 1, very short. Explosive capsule. *Mostly Old World*.

Placed with the Geraniales (p. 126) in most taxonomic systems, this is a family of 2 genera and 850 species. The largest genus is *Impatiens*, which is native to both Europe and North America, and of which some 20 species are cultivated.

CELASTRALES

Woody; petals usually present, free; disc often well developed.

112 **Cyrillaceae**. Woody. Leaves alternate, simple, exstipulate. Inflorescences racemes. Flowers bisexual, actinomorphic. KCA hypogynous, disc 0. K(5), C5/(5), A5/10, G(2–4); ovules 1–2 per cell, axile; style 1, short, stigmas 2–4. Fruit dry, indehiscent. *Mainly tropical America*.

A small family of 3 genera and 14 species; two of the genera are native to North America, and species of *Cyrilla* may occasionally be seen in European gardens.

113 **Aquifoliaceae**. Woody. Leaves alternate, simple, stipules small and falling early; ptyxis supervolute. Flowers in cymes/clusters, unisexual/rarely bisexual, actinomorphic. KCA hypogynous, disc 0. K4–5, C4–5/(4–5), A4–5 sometimes attached to base of C, G(3–n); ovules 1–2 per cell, axile; style 0, stigmas sessile. Drupe. *Widespread*.

Two genera (sometimes considered as only 1, *Ilex*) with about 400 species. *Ilex* is native to Europe and both *Ilex* and *Nemopanthus* are native to North America. Both genera are commonly cultivated.

114 **Corynocarpaceae**. Woody. Leaves alternate, stipulate, evergreen; ptyxis conduplicate. Flowers in panicles, bisexual, actinomorphic. KCA hypogynous. K5, C5, A5 on same radii as petals and attached to them at base, G(2); ovule 1, hanging; styles 1 or 2. Drupe. *Australasia, Polynesia*.

A family of a single genus (*Corynocarpus*) with 4 species, occasionally seen in cultivation.

115 **Celastraceae**. Woody, sometimes climbing. Leaves alternate/ opposite, simple, stipulate/exstipulate; ptyxis mainly involute. Inflorescences racemes/panicles, rarely flowers solitary. Flowers unisexual/bisexual, actinomorphic. KCA hypogynous/perigynous, with disc. K4–5, C4–5, A4–5, G(2–5); ovules usually 2 per cell, axile; styles 1 or as many as ovary cells. Fruit a capsule/berry/drupe/ samara, seeds with arils. *Widespread*.

A rather uniform, mainly tropical family of 70 genera and perhaps 1,300 species. Two genera are native to Europe and 13 to North America. Species of about 7 genera are grown as ornamental trees and shrubs; the most commonly seen is *Euonymus*.

116 **Staphyleaceae**. Trees/shrubs. Leaves usually opposite, compound/rarely simple, stipulate (stipules falling early); ptyxis involute. Inflorescences drooping panicles/racemes. KCA perigynous, with disc. K(5), C5, A5, G(2–4); ovules n, axile; styles 2–4, free. Inflated capsule. *N temperate area, S America, Asia*.

There are 5 genera and about 60 species. One genus occurs in Europe, 2 in North America. A few species of *Staphylea* are cultivated as flowering shrubs with curious, inflated and bladdery fruits.

117 **Stackhousiaceae**. Herbs with rhizomes. Leaves alternate, simple, exstipulate. Inflorescences spikes/racemes/clusters. Flowers bisexual, actinomorphic. KCA perigynous. K(5), C5, lobes either entirely free or free at base and apex, sometimes united in the middle, A5, G(2–5); ovules 1 per cell, axile/basal; style 1, divided into 2–5 segments at about half its length. Schizocarp. *Australasia*.

A uniform family of 2 or 3 genera and 38 species. One or 2 species have been cultivated in Europe, but are not commonly seen in gardens.

118 **Buxaceae**. Evergreen, usually woody. Leaves alternate/ opposite, simple, exstipulate. Flowers usually unisexual, actinomorphic. PA hypogynous. P4–6, A4, G(2–3); ovules 1–2 per cell,

axile; styles free. Capsule/berry-like, seeds shiny black. *Mostly tropics, some in temperate areas*.

Another rather uniform family of 5 genera and some 60 species. One genus is native to Europe, 3 to North America. Species of 4 of them are cultivated in gardens as evergreen shrubs. *Buxus* is also cultivated for its hard, smooth wood, and *Simmondsia* yields jojoba oil.

119 **Icacinaceae**. Trees/shrubs/woody climbers. Leaves alternate, simple, exstipulate; ptyxis conduplicate. Inflorescence a panicle. Flowers unisexual/bisexual, actinomorphic. KCA hypogynous. K(4–5), C4–5, A5, G(2–3); ovules 2 per cell, axile, pendulous; style 1, short, divided above. Drupe/samaras. *Tropics*.

A family of 60 genera and 320 species. Two of the genera are native in North America, and species of a few of them (*Apodytes, Villaresia*) occur occasionally in European gardens.

RHAMNALES

Woody; A on same radii as petals; KCA often perigynous, disc usually present; ovules 1–2 per cell.

120 **Rhamnaceae**. Woody. Leaves usually alternate and stipulate, simple; ptyxis conduplicate or involute (important in the classification of the family). Inflorescences corymbs/cymes/clusters. Flowers unisexual/bisexual, actinomorphic. KCA perigynous/epigynous, disc usually present. K4–5, C4–5, A4–5, on same radii as petals, G(2–4); ovules 1/rarely 2 per cell, axile; styles joined in the basal half. Capsule/drupe-like. *Tropics, N temperate areas*.

There are 53 genera and 875 species. Four genera are native to Europe and 14 to North America. Species of several genera are grown as ornamental shrubs or small trees.

121 **Vitaceae**. Usually woody climbers with tendrils/rarely shrubs/trees. Leaves alternate, simple/compound, stipulate/exstipulate; ptyxis conduplicate. Flowers small, in cymes, unisexual/bisexual, actinomorphic. KCA perigynous, disc present. K4–6/(4–6), C4–6/

(4–6) when united usually falling as a unit, A4–6 on the same radii as petals, G(2–6); ovules 1–2 per cell, axile. Berry. *Tropics & warm temperate areas.*

A family mainly of climbers in 13 genera and about 800 species. Two genera (1 introduced) occur in Europe and 4 in North America. About 8 genera are cultivated, some of them (*Ampelopsis*, *Cissus*, *Parthenocissus* and *Vitis*) being important climbers, others (*Cissus*, *Rhoicissus*) containing widely grown house-plants. The grape is produced by *Vitis vinifera*, native to the Mediterranean area.

122 Leeaceae. Shrubs, rarely scrambling/rarely herbaceous. Leaves alternate, simple/compound, exstipulate (though stalk-bases swollen). Flowers in cymes/panicles, bisexual, actinomorphic. KCA perigynous, with disc. K4–5/(4–5), C4–5, A4–5 on same radii as petals, G(4–8); ovules 1 per cell, axile; style 1. Berry. *Old World tropics.*

One genus (*Leea*) with about 34 species, a few of which are occasionally cultivated in glasshouses in Europe.

MALVALES

Stellate hairs common; K valvate in bud; filaments often united into a tube around ovary/style.

123 Elaeocarpaceae. Woody. Leaves alternate/opposite/rarely whorled, stipulate, deciduous or evergreen; ptyxis variable. Flowers in cymes/racemes/panicles/clusters, unisexual/bisexual, actinomorphic. KCA hypogynous. K4–5, C0/1–5, A n, G(2–5); ovules n, axile; style 1, lobed at the apex. Capsule/drupe-like. *Tropics.*

A mainly southern hemisphere family of 11 genera and 220 species. Three genera occur in North America, and species of 4 are cultivated as ornamental trees and shrubs.

124 Tiliaceae. Trees/shrubs/rarely herbs, often with stellate hairs. Leaves alternate, simple, evergreen/deciduous, stipulate (stipules often falling early); ptyxis conduplicate or supervolute. Flowers in cymes, bisexual, actinomorphic. KCA hypogynous. K4–5/(4–5),

C4–5, sometimes with nectaries on the petal-claws, A n, sometimes in 5 bundles, and/or the outermost sterile, G(3–5); ovules 1–n per cell, axile; style 1, very shortly lobed at top. Capsule/drupe/indehiscent. *Widespread*.

There are 48 genera and 725 species, 1 genus native to Europe, 3 to North America. Species of about 5 genera are cultivated, those of *Tilia* often forming important avenue- or street-trees.

125 Malvaceae. Herbs/woody, often with stellate hairs. Leaves alternate, simple/divided, stipulate; ptyxis variable. Inflorescences various. Flowers usually bisexual, actinomorphic. K hypogynous, CA perigynous. K5/(5) often with epicalyx, often with nectary-patch on inner surface, C5, contorted, free but all united at base to staminal tube, A (5–n) anthers 1-celled, G(2–n); ovules 1–n per cell, axile; style 1, divided in upper third or less. Capsule/schizocarp/berry. *Widespread*.

A large and usually easily-recognised family of 121 genera and 1,550 species. Fourteen genera (4 of them introduced) occur in Europe and 41 in North America. Species of about 25 genera are cultivated for ornament, and species of *Gossypium* provide the cotton of commerce.

126 Bombacaceae. Trees often with swollen trunks. Leaves simple/palmate, often scaly, stipules deciduous; ptyxis conduplicate, rarely conduplicate-plicate. Flowers large, bisexual, actinomorphic. KCA hypogynous/CA perigynous. K5/(5), C5 crumpled in bud, A5–n/(5–n) anthers 1-celled, G(2–5); ovules 2–n per cell, axile; style 1, stigmas capitate. Capsule/indehiscent, seeds often embedded in wool. *Tropics*.

Thirty genera with about 250 species. There are 4 genera native to the southern parts of North America, and species of a few are cultivated as ornamental glasshouse trees.

127 Sterculiaceae. Usually woody, often with stellate hairs. Leaves alternate, simple/compound, stipulate; ptyxis flat, conduplicate or conduplicate-plicate. Flowers solitary or clustered, unisexual/bisexual, actinomorphic. KCA hypogynous. K(5), C5/rarely 0, A(5)/(10), rarely free, G(2–5); ovules 2–n per cell, axile; styles 2–5, free/ 1 lobed at apex. Fruits various, sometimes splitting into apparently free carpels (follicles). *Mostly tropics*.

There are 73 genera and 1,500 species. Eleven genera are native to North America, and species of 13 genera are grown, mostly as greenhouse trees or shrubs.

THYMELEALES

Woody; leaves simple, exstipulate; C usually 0; G1.

128 Thymeleaceae. Usually woody. Leaves alternate/opposite, simple, exstipulate; ptyxis conduplicate or supervolute. Flowers in heads/racemes/spikes/rarely solitary, usually bisexual, actino-morphic. PA perigynous/rarely KCA perigynous. P(4–6), often tubu-lar, coloured/rarely K4, C4, small, A2–8, G1; ovules 1–2, more or less apical; style 1. Drupe/nut/capsule. *Widespread*.

A family of about 50 genera and 750 species. Three genera are native to Europe and 6 to North America. Species of about 10 genera are cultivated as ornamental shrubs, the most important genus being *Daphne*.

129 Elaeagnaceae. Woody, with conspicuous silvery or brown scales. Leaves alternate/opposite, entire, deciduous or evergreen, exstipulate; ptyxis variable. Flowers solitary or in clusters/racemes/ spikes, unisexual/bisexual, actinomorphic. PA perigynous. P(2–6), A4/8, G1; ovule 1, basal; style 1. Achene in persistent, fleshy peri-anth. *Widespread*.

There are 3 genera with 45 species. Two genera are native to Europe and 3 to North America. Species of all 3 genera are grown as shrubs with ornamental foliage, species of *Elaeagnus* being most commonly seen.

VIOLALES

Woody/herbaceous; C usually free; G superior/inferior; ovules usually parietal.

130 Flacourtiaceae. Shrubs/trees/woody climbers, sometimes spiny. Leaves alternate, simple, stipulate; ptyxis mainly supervolute. Inflorescence racemes/panicles/axillary clusters. Flowers unisexual/bisexual, actinomorphic. PA/KCA hypogynous/epigynous in some non-cultivated genera. C0 in all cultivated genera; P4–5/3+6, A4–5/7–10/n, G(2–5); ovules n, parietal; style 1/styles free/stigmas sessile. Capsule/berry. *Mostly tropics*.

A large family, with about 90 genera and 1,250 species. Nine genera are native in North America, and species of about 6 are occasionally seen in European gardens, the most frequent being *Azara*.

131 Violaceae. Herbs/shrubs. Leaves alternate/basal, stipulate, simple or divided; ptyxis involute. Flowers solitary/in clusters, actinomorphic/zygomorphic, usually bisexual. KCA hypogynous. K5/(5) often persistent, C5, A5 connectives appendaged, G(2–3); ovules 1–n per cell, parietal; style 1 or styles almost free, divided above, sometimes 2-, 3- or 6-fid. Capsule/berry. *Widespread*.

A family of about 20 genera and almost 1,000 species. *Viola* is native to Europe, and there are 4 genera (including *Viola*) native to North America. Species of *Viola* are grown as alpines and as bedding plants (pansies), and species of *Hymenanthera* are grown as ornamental shrubs.

132 Stachyuraceae. Shrubs or small trees. Leaves alternate, simple, stipulate (stipules falling early); ptyxis involute. Inflorescence racemose. Flowers usually bisexual, actinomorphic. KCA hypogynous. K4, C4, A8, G(4); ovules n, axile; style 1. Berry. *E Asia*.

A single genus (*Stachyurus*) with 5 or 6 species, a few of which are grown as ornamental shrubs.

133 Turneraceae. Shrubs/herbs. Leaves alternate, simple/lobed, exstipulate, often with 2 glands at base of blade. Flowers solitary or in clusters, flower-stalk sometimes partially fused with the subtending leaf-stalk, bisexual, actinomorphic. KCA perigynous/half epigynous. K(5)/5, C5, A5 G(3); ovules n, parietal; styles 3, free, broad, stigmas brush-like. Capsule, seeds with arils. *Mainly tropical America*.

A family of 10 genera and about 100 species. Two genera are native to north America, and 1 species of *Turnera* is occasionally grown as a glasshouse ornamental in Europe. Species of this genus are remarkable in that the flowers appear to be borne on the leaf-stalks (presumably by fusion of the flower- and leaf-stalks).

134 **Passifloraceae**. Trees/shrubs/climbers with tendrils. Leaves alternate, simple/compound, stipulate, stalks often with nectaries; ptyxis conduplicate. Flowers axillary, usually bisexual, actino-morphic. K & C united below, A & G often borne on a stalk (androgynophore). K3–8 usually 5, C usually 5/rarely 0, often with corona, A4–5/(4–5), G(3–5); ovules n, parietal; styles 3–5, united only at the base. Berry/capsule. *Mainly tropical America*.

There are 18 genera and 150 species. One genus is native to North America, and many species of *Passiflora* are grown as ornamental climbers (some hardy, others requiring glasshouse conditions).

135 **Cistaceae**. Herbs/shrubs. Leaves usually opposite, simple, usu-ally evergreen, stipulate/exstipulate; ptyxis conduplicate or rarely flat. Inflorescence usually cymose, rarely flowers solitary/in racemes. Flowers bisexual, actinomorphic. KCA hypogynous. K3–5 often dif-fering in size and shape, C3–5/rarely more, A n, G(3–10) usually (5); ovules 2–n, parietal; style 1 or stigmas head-like and sessile. *Mainly warm N temperate areas*.

A family of 7 genera and about 175 species of shrubs or shrub-lets. Five genera are native to Europe and 4 to North America. Spec-ies of a few genera (*Cistus*, *Fumana*, *Helianthemum*) are cultivated as hardy ornamental shrubs or shrublets.

136 **Bixaceae**. Trees/shrubs, sap red, orange or golden. Leaves alter-nate, simple, palmately veined/lobed, stipulate; ptyxis conduplicate. Inflorescences racemes/panicles. KCA hypogynous. K5, imbricate; C5; A n, G(2–5); ovules n, parietal; style 1. Capsule. *Tropics*.

There are 3 genera and 16 species. All 3 genera are native to North America, and species of all 3 are occasionally grown as ornamentals. *Bixa orellana* produces the red dye anatto, used in colouring foods.

137 **Tamaricaceae**. Shrubs/small trees. Leaves usually alternate, usually scale-like and with salt-secreting glands, exstipulate. Flowers in racemes/panicles/rarely solitary, bisexual, actinomorphic. KCA hypogynous. K4–6, C4–6, A4–15 arising from a disc, G(2–5); ovules n, parietal/basal. Capsule, seeds bearded at one end. *Mostly Mediterranean area & C Asia.*

A family found mainly in rather arid areas, consisting of 5 genera and about 90 species. Two genera are native to Europe and 1 to North America. Species of 3 genera are grown as ornamentals; the most commonly seen is *Tamarix,* which has leaves reduced to scales and pink or white flowers.

138 **Frankeniaceae**. Herbs/shrubs. Leaves opposite, entire, exstipulate, often with salt-secreting glands. Inflorescences cymes/flowers solitary. KCA hypogynous. K(4–7), C4–7 each with an outgrowth (ligule) on its inner face, A4–7/rarely more, usually 6, G(2–4); ovules 2–n per cell, parietal; style 1, divided above into 2–4 stigmas. Capsule. *Widespread.*

Found mainly in saline flats, there are 3 genera with 30 species. One genus (*Frankenia*) is native to both Europe and North America, and a few of its species are occasionally cultivated as ornamentals.

139 **Elatinaceae**. Small herbs, usually aquatic or semi-aquatic. Leaves opposite/whorled, stipulate, simple. Flowers solitary/inflorescences cymose, bisexual, actinomorphic. KCA hypogynous. K2–5 usually 4, C2–5 usually 4, A6–10 usually 8, G(2–5) usually (4); ovules n, axile; styles 2–5, usually 4, very small, free. Capsule. Seeds pitted. *Widespread.*

Two genera and 32 species of aquatic herbs. Both genera are native to Europe and North America, but only 1 (*Elatine*) is grown as an ornamental.

140 **Caricaceae**. Soft-wooded trees/shrubs. Leaves alternate, long-stalked, usually divided, exstipulate/stipules spine-like. Flowers

solitary/in cymes, usually unisexual, actinomorphic. KCA hypogynous. K(5), C(5), A10 attached to base of C, G(5); ovules n, parietal; styles 5, free. Large berry. *Tropical America & Africa*.

There are 4 genera and 31 species. Species of *Carica* are native to North America, and a few of them are also cultivated in Europe as glasshouse trees. *Carica papaya* produces the pawpaw of commerce.

141 **Loasaceae**. Herbs/shrubs, often with rough/stinging hairs. Leaves alternate/opposite, simple/divided, exstipulate. Flowers solitary and axillary or in cymes, bisexual, actinomorphic. KCA epigynous. K5, C5, A n, often united or united in bundles on same radii as petals, G(3–5); ovules n, parietal; style 1. Capsule. *Mainly America*.

There are 15 genera and 250 species; 4 of the genera are native to North America. A few species are cultivated as ornamental herbaceous plants.

142 **Datiscaceae**. Herbs/woody. Leaves alternate, simple/compound, exstipulate. Inflorescence raceme-like. Flowers unisexual, actinomorphic. PA/KCA epigynous. P3–8/rarely K3–8/(3–8), C0–8, A4–n, G(3); ovules n, parietal; styles free, sometimes bifid. Capsule. *Scattered*.

A diverse small family, with 3 genera and 4 species. One genus is native to Europe and 1 to North America. A single species (*Datisca cannabina*) is occasionally grown as an ornamental.

143 **Begoniaceae**. Herbs/shrubs. Leaves alternate, simple/compound, stipulate, often fleshy, base usually oblique; ptyxis conduplicate or conduplicate-plicate. Flowers usually unisexual, actinomorphic/zygomorphic. PA epigynous. P2–12 usually 4 in 2 pairs in male flowers, those of one pair often much smaller than those of the other pair, A n/(n), G(2–5) usually (3); ovules n, parietal/axile; styles 3, simple or divided. Capsule/berry. *Mostly tropics*.

A uniform family of 2 genera and almost 1,000 species. Both genera are native in the southern parts of north America, and species of both (especially *Begonia*) are grown as glasshouse plants, houseplants and half-hardy bedding annuals.

CUCURBITALES

Herbs, often with tendrils; fruits often large.

144 Cucurbitaceae. Mostly herbs/climbers with tendrils. Leaves alternate, often lobed, exstipulate; ptyxis conduplicate. Inflorescences axillary cymes/flowers solitary. Flowers usually unisexual, actinomorphic. KCA epigynous. K5/(5), C5/(5), A1–5 usually 3/ rarely (3), 1 anther 1-celled, G(3–5); ovules n, parietal/rarely axile; style 1, stigmas divided, sometimes complex, at apex/rarely styles 3, free. Fruit berry-like. *Widespread but mainly tropical.*

A distinctive family of 131 genera and 735 species. Seven genera (4 introduced) occur in Europe, and 26 genera are native to North America. About 28 genera are cultivated; several of them include economically important plants such as *Cucumis* (cucumber, melons) and *Cucurbita* (gourds, marrows, pumpkins).

MYRTALES

A n–few; G usually inferior; ovules axile/apical.

145 Lythraceae. Herbs, shrubs or trees. Leaves opposite/whorled, rarely alternate, stipulate/exstipulate; ptyxis flat or conduplicate. Inflorescences clusters/racemes/panicles. Flowers bisexual, actinomorphic/rarely zygomorphic, often heterostylous. KCA perigynous. K4–6 (epicalyx frequent), C2–6/rarely 0, A6–16, rarely fewer or more, G(2–6); ovules n, axile; style 1, stigma capitate. Capsule. *Widespread.*

Twenty-six genera and 580 species. Three genera occur in Europe and 11 are native in North America. Species of 5 or 6 genera are grown as ornamentals, some hardy (e.g. *Lythrum*), others requiring glasshouse protection (e.g. *Cuphea*).

146 Trapaceae. Aquatic annual herb. Leaves opposite below, alternate above, simple, exstipulate, with inflated stalks. Flowers solitary, actinomorphic, bisexual. KCA more or less epigynous. K4, C4,

A4, G(2); ovules 1 per cell, axile; style 1. Drupe with spines. *Old World.*

A single genus and about 30 species of aquatic herbs; *Trapa* is native to Europe. *T. natans*, the water-chestnut, is grown as a curiosity and is cultivated for its edible fruit in other parts of the world.

147 **Myrtaceae**. Woody. Leaves usually opposite, simple, exstipulate, with translucent aromatic glands; ptyxis flat, conduplicate or supervolute. Inflorescences various. Flowers bisexual, actinomorphic. KCA usually epigynous. K4–5/(4–5), C4–5/(4–5), A n/(n), G(3–n); ovules 2–n per cell, axile/parietal; style 1, long, slightly lobed at apex. Capsule/berry. *Mostly tropical America & Australia.*

A large and rather uniform family of 121 genera and 3,850 species. Two genera (1 introduced) occur in Europe and 17 are native in North America. Thirty-two genera are cultivated for ornament, and one of them, *Eucalyptus* is widely used as a rapidly-growing tree in southern Europe.

148 **Punicaceae**. Shrub/small tree, spiny. Leaves opposite or almost so, simple, exstipulate; ptyxis flat. Inflorescence cymose/flowers solitary. Flowers bisexual, actinomorphic. KCA epigynous. K(5–7), C5–7, A n, G usually (8–12), cells superposed/rarely (3); ovules n, axile; style 1. Fruit berry-like. *Temperate Old World.*

A single genus with 2 species, 1 of them native in Europe. One species (*Punica granatum*) is widely grown as an ornamental and for its edible fruit (pomegranate).

149 **Lecythidaceae**. Woody. Leaves alternate, simple, usually exstipulate; ptyxis supervolute (*Napoleona*). Flowers in large spikes, bisexual, actinomorphic/zygomorphic. KCA epigynous. K2–6, C4–8/rarely 0, A n variously united, G(2–6); ovules 1–n per cell, axile; style 1, short, lobed at apex. Fruit leathery or woody, seeds large and woody. *Tropics.*

A family of 20 genera and about 280 species. Species of a few genera are cultivated as ornamentals in glasshouses. *Bertholletia excelsa* produces the Brazil nut.

[145]

150 **Melastomataceae**. Woody/herbaceous. Leaves opposite/rarely whorled, exstipulate, simple, usually with 3 more or less parallel veins arising from base or near it; ptyxis conduplicate or supervolute. Inflorescences usually cymes. Flowers bisexual, perianth actinomorphic. KCA perigynous/epigynous. K3–6, C3–8, A3–n, usually 8/10, often unequal, filaments usually with a conspicuous joint, anthers opening by pores, G(2–5); ovules n, axile; style 1. Capsule/berry. *Mainly tropics*.

A large and relatively uniform family of 185 genera and over 4,000 species. Nineteen genera are native to north America, and species of over 30 are cultivated as glasshouse ornamentals.

151 **Rhizophoraceae**. Trees or shrubs (mangroves). Leaves alternate/opposite, simple, stipulate (stipules soon falling); ptyxis involute (*Crossostylis*). Inflorescences umbel-like/flowers solitary. Flowers usually bisexual, actinomorphic. KCA epigynous. K usually 4–5, C usually 4–5, A8–16/n, G(2–6); ovules 1 per cell, axile; style 1, short, lobed at tip. Fruit berry-like, seeds often partly developing while still on parent plant. *Tropics*.

Rarely seen in gardens except as uncharacteristic juvenile plants. There are 16 genera and about 130 species.

152 **Combretaceae**. Woody, sometimes climbing or with spines. Leaves alternate/opposite, simple, exstipulate, usually evergreen; ptyxis conduplicate or supervolute. Inflorescences spikes/racemes/panicles. Flowers bisexual/bisexual and male, actinomorphic. KCA epigynous. K4–5, C0/4–5, A8–10, G(2–5), 1-celled; ovules 2–6, apical; style 1, stigma capitate. Fruit 1-seeded, indehiscent, often winged or ridged. *Tropics*.

Nineteen genera and about 500 species. Five genera are native to North America, and species of about 4 are rarely grown as ornamental shrubs or climbers in glasshouses.

153 **Onagraceae**. Usually herbs. Leaves alternate/opposite, simple, stipulate/exstipulate; ptyxis flat or involute. Flowers solitary/racemose, bisexual, usually actinomorphic. KCA usually epigynous; tubular epigynous zone often present. K2–6 usually 4, C2–4/rarely 0,

A4–8/rarely 1, G(1–5) usually (4); ovules n, axile; style 1, lobed in upper third. Capsule/berry/nut. *Mostly temperate areas.*

There are 24 genera and 650 species. Five genera (1 introduced) occur in Europe, and 13 in North America. Species of about 10 genera are grown as ornamentals, the most important being *Fuchsia*, which includes both hardy and glasshouse species.

154 Haloragaceae. Herbs, aquatic or in moist habitats. Leaves alternate/opposite/whorled/all basal, stipulate/exstipulate; ptyxis flat or conduplicate-plicate. Inflorescences various. Flowers unisexual/bisexual, actinomorphic. KCA/PA epigynous. K0–(4), C0–4, A8, G(1–4); ovules 1/1 per cell, axile/apical; styles 1–4, free. Nut/drupe. *Widespread.*

In the broad sense, there are about 10 genera and 160 species. It is sometimes divided into 2 families.

1a Ovary 2–4-celled; stipules absent; leaves deeply dissected; stamens 4 or more   **154a Haloragaceae**
 b Ovary 1-celled; stipules present; leaves entire or lobed; stamens 1 or 2   **154b Gunneraceae**

154a Haloragaceae in the strict sense. Mostly aquatic herbs/rarely shrubs/small trees. Leaves deeply dissected, without stipules. Flowers solitary/in terminal spikes/panicles, unisexual/bisexual; stamens 4 or more; ovary 2–4-celled. *Widespread.*

In the narrow sense, there are 9 genera and about 120 species. One genus occurs in Europe and 5 in North America. A few species of *Myriophyllum* are grown as aquarium plants.

154b Gunneraceae. Terrestrial herbs but often growing in marshy places or at water margins. Leaves often large and rhubarb-like, not divided, stipules present. Flowers in panicles, unisexual; stamens 1 or 2; ovary 1-celled. *S hemisphere.*

The single genus is *Gunnera*; several species are cultivated as spectacular pond-margin plants.

155 **Theligonaceae**. Fleshy herb. Lower leaves opposite, upper alternate, all simple and with sheathing bases. Inflorescence cymose. Flowers unisexual; male: actinomorphic, PA hypogynous, P2, A7–22; female: more or less zygomorphic, P tubular, G1, style 1, at last lateral; ovule 1, basal. Nut. *Scattered*.

A single genus with 3 species; the genus occurs wild in Europe.

156 **Hippuridaceae**. Aquatic rhizomatous herb. Leaves whorled, entire, exstipulate. Flowers axillary, unisexual/bisexual, actinomorphic. P0, A1 epigynous, G1; ovule 1, apical; style 1. Cypsela. *Temperate areas*.

A single genus with a single species, native in Europe and North America, and grown as an aquarium and pond plant.

157 **Cynomoriaceae**. Root parasites without chlorophyll. Leaves scale-like. Inflorescence spike-like or head-like. Flowers usually unisexual, actinomorphic. PA epigynous. P1–5, A1, G1; ovule 1, more or less apical. Small nut. *Mediterranean area to C Asia*.

A single genus, native to Europe, with 2 species.

UMBELLALES

Flowers with C free, usually in umbels; G inferior; ovules axile.

158 **Alangiaceae**. Trees/shrubs, sometimes spiny, rarely lianas, with latex. Leaves alternate, simple/lobed, exstipulate. Inflorescences axillary cymes. Flowers bisexual, actinomorphic. KCA epigynous. K(4–10) rarely almost 0, C4–10, recurving, A4–20, G(2–3), 1-celled; ovules 1 per cell, pendulous; style 1, capitate. Drupe. *Old World tropics & subtropics*.

There is a single genus (*Alangium*) with about 18 species. One or 2 species are occasionally grown as specimen trees.

159 **Nyssaceae**. Trees/shrubs. Leaves alternate, simple, exstipulate; ptyxis conduplicate. Flowers in axillary clusters, unisexual/bisexual, actinomorphic. KCA epigynous. K5, C5, A8–12, G(2), 1-celled;

ovule 1, axile; style 1, lobed at apex. Drupe. *Temperate N America, China*.

There are 2 genera and 7 species; 1 genus is native in north America, and species of both are cultivated, especially those of *Nyssa*, which often show spectacular autumn coloration.

160 **Davidiaceae**. Trees. Leaves alternate, simple, exstipulate. Inflorescence a head surrounded by 2 showy white bracts, consisting of 1 bisexual flower surrounded by many male flowers. P0, A1–7, epigynous in bisexual flowers, G6–10-celled; ovules 1 per cell, axile; style 1. Drupe. *China*.

A single genus (*Davidia*) with a single species, sometimes included in the *Nyssaceae*. *D. involucrata*, the handkerchief tree, is widely grown as an ornamental tree.

161 **Cornaceae**. Usually woody. Leaves usually opposite/rarely alternate, simple, exstipulate, sometimes evergreen; ptyxis conduplicate or involute. Flowers in corymbs or umbels, unisexual/bisexual, actinomorphic. KCA epigynous, K3–5, C3–5/rarely 0, A3–5, G(2–3); ovules 1 per cell, axile/rarely parietal; style 1, slightly lobed above, or styles 2–3, free. Drupe/berry. *Mostly temperate areas*.

A family of 13 genera and over 100 species. One genus (*Cornus*) is native to both Europe and north America. Species of about 5 genera are grown as ornamental shrubs or small trees.

162 **Garryaceae**. Evergreen shrubs/trees. Leaves opposite, entire, exstipulate; ptyxis conduplicate. Inflorescences catkin-like. Flowers unisexual; male: P4, A4; female: P0, G(2) naked/inferior; ovules 2, apical; styles 2, free, stigmas lateral. Berry. *N America*.

A single genus (*Garrya*), with about 8 species; the genus is native to western North America, and several of its species are grown as ornamental trees. The inflorescences technically qualify as catkins, but this family is not really similar to those others with catkinate inflorescences.

163 **Araliaceae**. Herbs/shrubs/trees, often spiny. Leaves alternate, usually lobed/compound, stipulate, stellate hairs frequent; ptyxis

mainly conduplicate. Inflorescences umbels, these often aggregated into complex panicles. Flowers unisexual/bisexual, actinomorphic. KCA epigynous. K5/(4–5)/rim-like, C4–15 usually 5, usually valvate, A4–15 usually 5, G(2–30); ovules 1 per cell, axile; style single or styles free and as many as ovary-cells. Berry/drupe. *Mostly tropics*.

A large and fairly uniform family of about 15 genera and over 1,400 species. One genus (*Hedera*) is native to Europe and there are 11 native to North America. Species of 18 genera are cultivated as evergreen climbers, herbs or shrubs. The distinctions between this family and the *Umbelliferae* (see below) become very tenuous in the tropics.

164 **Umbelliferae/Apiaceae**. Usually herbs. Leaves alternate, often pinnately compound, stalks sheathing; ptyxis conduplicate or supervolute. Inflorescences umbels/rarely heads. Flowers usually bisexual, actinomorphic/zygomorphic. KCA epigynous. K(5) often very reduced, C5 imbricate and inflexed, A5, G(2); ovules 1 per cell, axile; styles 2, free or slightly united at base. Schizocarp. *Widespread*.

A large and uniform family with 420 genera and 3,100 species. One hundred and ten genera are native to Europe, and 83 to North America, and species of very many genera are cultivated, either as ornamentals, as vegetables — carrot (*Daucus carota*), celery (*Apium graveolens* var. *dulce*), parsnip (*Pastinaca sativa*) or as herbs — caraway (*Carum carvi*), coriander (*Coriandrum sativum*), cumin (*Cuminum cyminum*), etc.

DIAPENSIALES

Shrublets; 5 staminodes present; A opening by slits.

165 **Diapensiaceae**. Evergreen herbs/shrublets. Leaves alternate, simple, exstipulate; ptyxis conduplicate. Flowers solitary or in heads/racemes, bisexual, actinomorphic. K hypogynous, CA perigynous. K(5), C(5), A5 with 5–0 staminodes, G(3); ovules usually n, axile; style 1, short, 3-lobed above. Capsule. *N temperate areas*.

A small family of 6 genera and 13 species, 1 genus native to

Europe, 4 to North America. Species of 3 of the genera are cultivated as ornamentals.

ERICALES

Usually woody; leaves simple, exstipulate, often evergreen; anthers often opening by pores; pollen often remaining in tetrads.

166 **Clethraceae**. Shrubs/trees, evergreen/deciduous. Leaves alternate, simple, exstipulate, with stellate hairs; ptyxis supervolute. Inflorescences racemes/panicles. Flowers fragrant, bisexual, actinomorphic, disc 0. KCA hypogynous. K(5–6), C5–6, A10–12, sometimes slightly attached to base of petals, G(3); ovules n per cell, axile; style 1, stigma 3-lobed. Capsule. *Mostly tropics & subtropics.*

There is a single genus (*Clethra*) with about 64 species, some of which are native to North America. About 7 species are cultivated as ornamental shrubs.

167 **Pyrolaceae**. Herbs/shrubs, sometimes saprophytic. Leaves alternate, often in rosettes, evergreen, exstipulate. Flowers solitary/in racemose inflorescences, bisexual, actinomorphic. KCA hypogynous. K4–5, C4–5, A8–10 anthers opening by pores, G(4–5); ovules n per cell, axile/parietal; style 1, stigmas divided at apex. Capsule/berry. *N temperate areas.*

Often included in *Ericaceae*, there are 5 genera, all found in Europe, and about 42 species. Several species in 3 of the genera are cultivated as ornamentals, even though they are often difficult to grow well.

168 **Ericaceae**. Woody/herbs/rarely saprophytic or parasitic and lacking chlorophyll. Leaves alternate/opposite/appearing whorled/basal, simple, exstipulate, usually evergreen, sometimes needle-like; ptyxis very variable (important in the classification of *Rhododendron*). Inflorescences racemes/clusters/flowers solitary. Flowers bisexual, actinomorphic/zygomorphic. KCA hypogynous/epigynous, rarely K hypogynous CA perigynous, disc present. K(4–5)/rarely 4–5, sometimes very small, C(3–5)/3–5/rarely –10, A5–10/rarely –27 anthers

usually opening by pores, pollen usually in tetrads, G(2–12); ovules n, axile/rarely parietal; style 1, lobed at apex, the stigmas often held within a sheath. Capsule/berry/drupe. *Widespread*.

A very large and diverse family, with over 100 genera and over 3,000 species. Eighteen genera occur in Europe and over 40 in north America. Species of many of the genera are cultivated, especially the two largest, *Erica* (heathers) and *Rhododendron*.

Formerly 3 segregate families were recognised within the broad *Ericaceae*: *Ericaceae* in the strict sense (plants with chlorophyll, ovary superior), *Vacciniaceae* (plants with chlorophyll, ovary inferior) and *Monotropaceae* (plants saprophytic/parasitic, without chlorophyll).

169   **Empetraceae**. Evergreen, heath-like shrublets. Leaves alternate/ almost whorled, entire, exstipulate. Flowers solitary/in terminal clusters, unisexual/bisexual, actinomorphic. PA hypogynous, disc 0. P2– 6, A3–4, G(2–9); ovules 1 per cell, axile; style 1 with 2–9 branches above. Drupe with 2–9 stones. *Temperate areas*.

Three genera with 5 species with a remarkable distribution in both north and south temperate areas. Two genera are native to Europe, 3 to north America. Species of 2 of them are occasionally cultivated.

170   **Epacridaceae**. Shrubs/small trees. Leaves alternate, rigid, simple, exstipulate; ptyxis flat. Inflorescences spikes/racemes/panicles. Flowers bisexual, actinomorphic. K hypogynous, CA perigynous. K(4–5), C(4–5), A4–5, anthers 1-celled, G(2–10); ovules 1–n per cell, axile; style 1. Capsule/drupe. *Mainly Australasia*.

A rather uniform family of 31 genera and 400 species, particularly well developed in Australia. Species of a few of the genera are cultivated.

PRIMULALES

Flowers usually actinomorphic; petals usually united; A on the same radii as the corolla-lobes; ovules free-central/basal.

[152]

171   **Theophrastaceae**. Shrubs. Leaves alternate, sometimes in false terminal whorls, simple, exstipulate; ptyxis conduplicate. Flowers solitary/paired/in racemes, usually bisexual, actinomorphic. K hypogynous, CA perigynous. K(4–6) usually 5, C(4–6) usually 5, A4–6 usually 5, on same radii as C-lobes, anthers opening by pores towards the outside of the flower, with usually 5 staminodes, G1-celled; ovules n, free-central; style simple, lobed at apex. Drupe. *New World tropics*.

There are 4 genera and 100 species; 1 genus occurs in north America. Species of all 4 genera are grown as ornamentals in glasshouses.

172   **Myrsinaceae**. Woody. Leaves alternate/rarely whorled, exstipulate, mostly evergreen, with translucent or coloured dots or stripes; ptyxis variable, mainly supervolute. Inflorescences cymose/clusters. Flowers unisexual/bisexual, usually actinomorphic. K hypogynous, CA perigynous/rarely KCA half-epigynous. K(4–5), C(4–5)/rarely 4–5, A4–5, on same radii as C-lobes, opening by slits towards inside of flower, G(4–5); ovules n, free-central, style 1, stigmas 4–5. Berry/drupe. *Mainly tropics*.

A uniform family of 39 genera and about 1,250 species. One genus is native to Europe, 8 to North America. Species of about 4 genera are occasionally grown as ornamental shrubs.

173   **Primulaceae**. Herbs/rarely shrublets/rarely aquatic. Leaves alternate/opposite/basal, usually simple, exstipulate; ptyxis variable (important in the classification of *Primula*). Inflorescence various, often superposed whorls. Flowers bisexual, actinomorphic/rarely zygomorphic. K hypogynous CA perigynous/rarely KCA half-epigynous/rarely PA hypogynous. K(5–7), C(5–7)/rarely 0, A5–7 on same radii as C-lobes, G usually (5), 1-celled; ovules usually n, free-central; style 1, stigma capitate. Capsule. *Widespread*.

A uniform family of 22 genera and about 800 species. Fourteen genera are native to Europe, 10 to North America. Many of the species are cultivated, especially those belonging to the largest genus, *Primula*.

PLUMBAGINALES

Stamens on same radii as petals; ovule 1, basal on long curved stalk.

174 **Plumbaginaceae**. Herbs/shrubs/climbers. Leaves alternate/ basal, simple, exstipulate; ptyxis flat or involute. Inflorescence racemose or cymose, flowers often aggregated into 'spikelets'. Flowers bisexual, actinomorphic. KCA hypogynous/K hypogynous CA perigynous. K5/(5), C(5)/rarely 5, A5 on same radii as C-lobes, G(5); ovule 1, basal; styles 5, free. Fruit indehiscent, retained in K-tube. *Widespread*.

A relatively uniform family, in which between 15 and 27 genera are recognised by different authors; there are over 400 species, many of them growing only in brackish marshes. Eight genera are native to Europe, 3 to North America. Species of about 5 genera are cultivated as ornamental herbaceous plants; several have 'everlasting' flowers and are grown for the cut-flower trade.

EBENALES

Woody; leaves simple; flowers actinomorphic; A usually more numerous than C-lobes.

175 **Sapotaceae**. Trees/shrubs/woody climbers, with milky sap, sometimes spiny. Leaves alternate/rarely opposite/whorled, simple, usually exstipulate, leathery; ptyxis conduplicate. Flowers solitary/ in clusters, actinomorphic, unisexual/bisexual. K hypogynous CA perigynous. K4–11/(4–6), C(4–18), A4–43 often some staminodial, G(1–30); ovules usually 1 per cell, style 1, stigma capitate or lobed. Berry/drupe. *Mostly tropics*.

A family of between 53 and 100 genera and 1,100 species. Seventeen genera are native to North America, and species of about 11 of the genera are occasionally cultivated as ornamental shrubs.

176 **Ebenaceae**. Woody, wood dark, sap watery. Leaves alternate, simple, entire, leathery, exstipulate; ptyxis conduplicate or super-

volute. Flowers solitary/in cymes. K hypogynous, CA perigynous, actinomorphic, usually unisexual. K(4), C(4), A 6–20, G(2–16); ovules 1–2 per cell, axile, pendulous; style 1, stigma capitate or style short, divided above into as many branches as there are ovary cells. Berry. *Mostly tropics*.

There are between 2 and 5 genera and about 500 species. One genus is introduced in Europe, and 1 is native in North America. A few species of *Diospyros* are grown as ornamental trees. Ebony is the wood from *D. ebenum* and perhaps other species.

177 **Styracaceae**. Trees/shrubs. Leaves alternate, simple, exstipulate, with stellate hairs or scales; ptyxis mainly supervolute. Inflorescences panicles/racemes/clusters. Flowers bisexual, actinomorphic. KCA hypogynous/epigynous/K hypogynous CA perigynous. K(2–10), C(4–8), A4–8/8–16, G(2–5); ovules n, axile; style 1, stigma capitate or 2–5-lobed. Drupe/capsule. *E Asia, America, Mediterranean area*.

A family of 12 genera and 165 species. One genus is native to Europe and 2 to north America. Species of 5 of the genera are cultivated as ornamental small trees.

178 **Symplocaceae**. Trees/shrubs. Leaves alternate, simple, exstipulate, leathery. Flowers in spikes/racemes/panicles/rarely solitary, bisexual, actinomorphic. KCA more or less epigynous. K(4–5), C(4–10), A4–n, G(2–5); ovules 2–4 per cell, axile; style 1. Berry/drupe. *Tropical America, Asia, Australasia*.

A single genus (*Symplocos*) with about 250 species; a few species are native to North America, and a few are grown as ornamentals (largely for their bluish fruits).

OLEALES

Woody; A2.

179 **Oleaceae**. Woody, sometimes climbing. Leaves usually opposite, simple/pinnately compound, exstipulate; ptyxis flat, conduplicate or supervolute. Inflorescence often a cymose panicle. Flowers

[155]

usually bisexual, actinomorphic. KCA hypogynous/K hypogynous CA perigynous. K(4)/rarely 0(–15). C(4)/rarely 0(–15), A2, G(2); ovules usually 2 per cell, axile; style 1, divided above into 2 stigmas. Fruit various. *Temperate areas & tropics*.

A relatively uniform family of 24 genera and over 900 species. Nine genera are native to Europe and 12 to North America. Many species of several genera (especially *Forsythia* and *Syringa*) are grown as ornamental shrubs. *Olea europaea* produces the olive.

GENTIANALES

Leaves usually entire; flowers usually actinomorphic; A4–5, G mostly (2), superior.

180 **Loganiaceae**. Woody, sometimes climbing, with internal phloem; glandular hairs absent. Leaves opposite, entire, stipulate; ptyxis flat. Inflorescences cymose/flowers solitary. Flowers bisexual, actinomorphic. K hypogynous CA perigynous. K(4–5) imbricate, C(4–5), A4–5, G(2); ovules n, axile; style 1, stigma capitate or bilobed. Capsule/berry/drupe. *Tropics*.

A mainly tropical family of 29 genera and 800 species. Five genera are found in North America, and species of a few are cultivated. The *Buddlejaceae* (see below) is sometimes included within this family.

181 **Desfontainiaceae**. Shrubs. Leaves opposite, exstipulate, evergreen, spine-margined; ptyxis conduplicate. Flowers solitary/inflorescences cymose, bisexual, actinomorphic. K hypogynous CA perigynous. K5 spine-like, C(5) with long tube, A5, G(5); ovules n, axile; style 1. Berry. *Andes*.

One genus and 1 species (*Desfontainia spinosa*) cultivated in European gardens for its holly-like foliage and large, tubular, red flowers.

182 **Gentianaceae**. Herbs, sometimes climbing/small shrubs. Leaves mostly opposite/rarely whorled/alternate, simple, exstipulate; ptyxis variable, often supervolute. Flowers solitary or in cymes/panicles, bisexual. K hypogynous CA perigynous. K(4–5) rarely –(12), C(4–

5) rarely —(12), A4–5 rarely -12 anthers opening by slits/rarely pores, G(2); ovules n, usually parietal; style 1 divided above, or stigmas 2, sessile. Capsule/berry-like. *Widespread.*

A uniform family of 80 genera and 700 species. Nine genera occur in Europe and 17 in North America. Species of 12 genera are cultivated, the most popular belonging to the largest genus, *Gentiana*.

183 **Menyanthaceae**. Aquatic/marsh herbs. Leaves alternate, entire/of 3 leaflets, stalks sheathing; ptyxis involute or supervolute. Inflorescences various. Flowers bisexual, actinomorphic. K hypogynous, CA perigynous. K(5)/5, C(5) valvate, A5, G(2); ovules n, parietal; style 1 bifid into 2 stigmas at apex. Fruit usually a capsule. *Temperate areas.*

Five genera with 40 species, sometimes included in the *Gentianaceae*. Two genera are native to Europe and 2 to North America; species of 2 are cultivated as ornamental aquatics.

184 **Apocynaceae**. Woody/herbs, often climbing, with milky sap. Leaves entire, usually opposite, exstipulate; ptyxis flat, conduplicate or involute. Inflorescences racemose/cymose/flowers solitary. Flowers bisexual, actinomorphic. K hypogynous CA perigynous. K4–5/(4–5), C(5) contorted, A5, G(2), often united only by the common style, which usually has a swelling below the stigmatic area; ovules n, marginal/axile. Fruit various, seeds often plumed. *Widespread, mainly centred in tropics.*

A rather uniform family of 215 genera and 2,100 species. Four genera are native to Europe, 27 to North America. Species of about 12 genera are cultivated as ornamentals.

185 **Asclepiadaceae**. Woody/herbs/climbers, usually with milky sap. Leaves opposite, entire, stipules minute/0; ptyxis flat. Inflorescences racemose/cymose/flowers solitary. Flowers bisexual, actinomorphic. K hypogynous CA perigynous. K(5)/5, C(5) contorted, corona frequent, A5 often joined to style, 'translators' and pollinia frequent, G(2), often united only by common style; ovules n, mar-

[157]

ginal/axile; style 1, large, head-like. Fruit of 1–2 follicles, seeds plumed. *Mostly tropics*.

Very similar to the *Apocynaceae* and occasionally included within it, this family includes 348 genera and 2,900 species. Eight genera are native to Europe, 15 to North America. Species of a wide range of genera are grown as ornamentals.

186 **Rubiaceae**. Herbs/shrubs/trees/rarely scramblers. Leaves opposite/whorled, stipulate, stipules sometimes leaf-like; ptyxis mainly flat, occasionally revolute or supervolute. Inflorescences various. Flowers usually bisexual, usually actinomorphic. KCA epigynous or half-epigynous. K(4–5) rarely –(12), C(4–5) rarely –(12), A3–5 rarely –12, G(2–5), ovaries of several flowers sometimes coalescing; ovules 1–n per cell, axile; style 1, divided above, or 2, free. Capsule/berry/schizocarp. *Widespread*.

A very large family, with 631 genera and 10,700 species. Nine genera are native to Europe and 60 to north America. Species of about 34 genera are cultivated, but for its size, the family contributes remarkably little to gardening.

SCROPHULARIALES

C often 2-lipped; A4/2; G(2), superior.

187 **Polemoniaceae**. Herbs/rarely shrubs/climbers with tendrils. Leaves alternate/opposite, entire/pinnately divided, usually exstipulate; ptyxis conduplicate. Inflorescence cymose to head-like/rarely flowers solitary. Flowers bisexual, usually actinomorphic. K hypogynous CA perigynous. K(5), C(5) contorted, A5, G(3); ovules 1–n per cell, axile; style 1 divided above into 3 stigmas. Capsule. *Mainly America*.

There are twenty genera with 275 species. Two genera (1 introduced) are found in Europe and 14 in North America. Species of about 8 genera are cultivated as ornamental herbaceous plants.

188 **Fouquieriaceae**. Woody, spiny. Leaves alternate, simple, exstipulate, fleshy. Inflorescence a terminal panicle. Flowers bisex-

ual, actinomorphic. K hypogynous CA perigynous. K5, C(5), A10–17, G(3); ovules 12–18, parietal; style 1, stigma 3-lobed at the apex. Capsule. *C & SW North America*.

One genus with 11 succulent species, most of them found in western North America, very rarely cultivated in Europe.

189 **Convolvulaceae**. Climbers/shrublets, often with milky sap/rarely twining parasites without chlorophyll. Leaves alternate, simple, exstipulate, scale-like in parasites; ptyxis conduplicate. Inflorescence cymose/clustered/flowers solitary. Flowers bisexual, actinomorphic. K hypogynous CA perigynous. K4–5/(4–5), C(4–5) contorted, A5 sometimes with scales below their insertion, G(2) sometimes 4-celled; ovules 1–2 per cell, axile/parietal; styles 2 free or united and stigma capitate. Capsule/fleshy. *Widespread*.

A relatively uniform family of 58 genera and 1,650 species. Six genera are native to Europe, 18 to north America. Species of about 8 genera are cultivated for ornament.

The parasitic genus *Cuscuta* is sometimes separated off into the family *Cuscutaceae*.

190 **Hydrophyllaceae**. Herbs/rarely shrubs occasionally with irritant or stinging hairs. Leaves alternate/basal/rarely opposite, entire/divided, exstipulate; ptyxis flat or conduplicate. Inflorescences coiled cymes/flowers solitary. Flowers bisexual, actinomorphic. K hypogynous CA perigynous. K5/(5), C(5) usually imbricate, A5, G(2); ovules n/rarely 4, parietal/rarely axile; style 1 or styles 2, free or variously united below. Capsule. *Mainly America*.

A smallish family, with 22 genera and 275 species. Seventeen genera are native to North America, and 2 of them are introduced in Europe. Species of 9 genera are cultivated for ornament, members of the genus *Phacelia* being most frequently seen.

191 **Boraginaceae**. Herbs/woody. Leaves alternate, simple, exstipulate; ptyxis conduplicate or supervolute. Inflorescences often coiled cymes. Flowers bisexual, actinomorphic/rarely zygomorphic. K hypogynous CA perigynous. K5/(5), C(5), A5, G(2) usually 4-celled by secondary septa; ovules 4, side-by-side, axile; style 1, terminal or

from between the 4 cells, stigma simple or slightly 2-lobed. Fruit 4/rarely 1 nutlets/drupe. *Widespread*.

A large family with 156 genera and 2,500 species. Thirty-four genera are native to Europe and 34 to north America. Species of many of them are cultivated as ornamentals.

192 **Lennoaceae**. Parasitic herbs, chlorophyll 0. Leaves scale-like. Inflorescence spicate/cymose/head-like. Flowers bisexual, actinomorphic. K hypogynous CA perigynous. K(6–10), C(5–8) imbricate, A5–8, G(6–15); ovules 2 per cell, axile; style 1. Fleshy capsule. *SW USA, Mexico*.

Three genera and 6 species of parasites, 2 of the genera native to North America.

193 **Verbenaceae**. Woody/herbaceous. Leaves opposite, simple/compound, exstipulate; ptyxis mainly conduplicate. Inflorescences various. Flowers bisexual, zygomorphic. K hypogynous CA perigynous. K(5–8) more or less actinomorphic, C(5), A4/rarely 2–5, G2–9-celled, style 1, terminal, divided into 2–9 stigmas at apex; ovules 1–2 per cell, axile/rarely parietal. Drupe/berry/rarely 4 nutlets. *Mainly tropics*.

There are 91 genera and 1,900 species. Four genera (1 introduced) occur in Europe and 23 in North America. Species of about 25 genera (especially *Verbena* and *Callicarpa*) are cultivated.

194 **Callitrichaceae**. Aquatic herbs. Leaves opposite, simple, exstipulate. Flowers solitary, axillary. unisexual, actinomorphic, A hypogynous. P0, A1, G(2), 4-celled by secondary septa; ovules 1 per cell, axile; styles 2, free. Schizocarp. *Widespread*.

The single genus (*Callitriche*) contains 17 species, and is found in both Europe and North America. A few species are cultivated as aquarium plants.

195 **Labiatae/Lamiaceae**. Herbs/shrubs. Leaves opposite, aromatic, simple/compound, exstipulate; ptyxis variable but mainly conduplicate. Flowers usually in verticils, mostly bisexual, zygomorphic. K hypogynous CA perigynous. K usually (5) often zygomorphic, C(5)/

rarely (3) 1–2-lipped, A4/2, G(2), 4-celled by secondary septa; style usually from between the 4 cells/rarely terminal, divided at the apex into 2 stigmas; ovules 1 per cell, axile. Fruit of 4 nutlets/rarely fleshy. *Widespread*.

A rather uniform family of 224 genera and 5,600 species. Forty-one genera are found in Europe and 70 in north America. Species of about 70 genera are cultivated as ornamentals or as aromatic herbs – mint (*Mentha*), rosemary (*Rosmarinus*), thyme (*Thymus*), etc.

196 **Nolanaceae**. Herbs/shrublets. Leaves alternate, simple, exstipulate, fleshy. Flowers axillary, bisexual, actinomorphic. K hypogynous CA perigynous. K(5), C(5) infolded in bud, A5, G(5), lobed; ovules few, axile; style 1, divided at the apex into 2 stigmas. Schizocarp. *Chile, Peru*.

There are two genera (sometimes combined into 1) with about 16 species, a few of which are cultivated for ornament.

197 **Solanaceae**. Woody/herbaceous, with internal phloem. Leaves alternate, simple/rarely pinnatisect, exstipulate; ptyxis conduplicate. Inflorescence often cymose/flowers solitary, often extra-axillary. Flowers bisexual, actinomorphic/zygomorphic. K hypogynous CA perigynous. K5/(5), C(5) lobes folded/contorted/valvate, A5/rarely 4/2, G usually (2), septum usually oblique to median plane of flower, rarely with secondary septa; ovules n, axile; style 1, stigma bilobed or capitate. Berry/capsule. *Widespread*.

A large and diverse family with 90 genera and 2,600 species. Fourteen genera (5 of them introduced) are found in Europe and 33 in north America. Species of about 32 genera are cultivated, either for ornament or as vegetables – aubergine (*Solanum melongena*), potato (*Solanum tuberosum*), tomato (*Lycopersicon esculentum*), etc.

198 **Buddlejaceae**. Woody/rarely herbaceous, without internal phloem, often with glandular hairs. Leaves opposite/whorled/rarely alternate, often toothed, stipules forming a line uniting the leaf-bases; ptyxis flat-conduplicate. Inflorescences various. Flowers bisexual, actinomorphic. K hypogynous CA perigynous. K(4), C(4), A4,

[161]

G(2), style 1, stigma capitate or 2-lobed; ovules n, axile. Capsule/ berry drupe. *Mainly tropical E Asia*.

A single genus with about 100 species; it is now introduced from cultivation in both Europe and North America and is often placed close to, or included in *Loganiaceae* (see above).

199 **Scrophulariaceae**. Herbs/woody, some half-parasitic; internal phloem absent. Leaves alternate/opposite, simple/rarely compound, exstipulate; ptyxis variable. Inflorescences various. Flowers bisexual, zygomorphic. K hypogynous CA perigynous. K(4–5), C(4–5)/ rarely –(8) imbricate, A4/2/rarely 5, G(2), septum at right angles to median plane of flower; ovules 1–n, axile; style 1, stigma capitate or bilobed. Capsule/rarely berry/indehiscent. *Widespread*.

There are over 220 genera and 4,500 species. Thirty-nine genera are native in Europe and 69 in North America. Species of about 70 genera are cultivated as ornamentals.

200 **Globulariaceae**. Herbs/shrublets. Leaves alternate/basal, ex-stipulate; ptyxis conduplicate. Inflorescence a bracteate head. Flowers bisexual, zygomorphic. K hypogynous CA perigynous. K(5), C(4–5), A4, G(2), 1-celled; ovule 1, apical; style 1, stigma capitate. Nut. *Mostly Mediterranean area*.

A family considered to contain only a single genus (*Globularia*), or by some to include other, mainly South African genera (*Hebenstreitia*, *Selago*, etc.) which are often placed in the *Scrophulariaceae* or in a distinct family, *Selaginaceae*. *Globularia*, *Hebestreitia* and *Selago* are all cultivated in Europe, but only the first is commonly found.

201 **Bignoniaceae**. Usually woody and climbing, often with leaf-tendrils/rarely herbs. Leaves usually opposite, compound/rarely simple; ptyxis conduplicate. Inflorescence usually cymose. Flowers bisexual, zygomorphic. K hypogynous CA perigynous. K(5), C(5), A4/rarely 2, G(2); ovules n, axile/rarely parietal; style 1, stigma capitate or bilobed. Fruit usually a capsule. Seeds often winged. *Mainly tropics*.

The distinction between this family and the *Scrophulariaceae* is sometimes difficult. There are 112 genera and 725 species; 18 genera are native to North America. Species of 28 genera are grown, as spectacular ornamental climbers or trees (e.g. *Campsis*, *Bignonia*, *Jacaranda*, etc.).

202 **Acanthaceae**. Usually herbs. Leaves opposite, simple, exstipulate, often with cystoliths; ptyxis variable. Inflorescences cymose, often with conspicuous overlapping bracts. Flowers bisexual, zygomorphic. K hypogynous CA perigynous. K(4–5), C(5) 2-lipped, A4/2, G(2); ovules axile, 2 or more per cell; style 1, slightly bilobed at apex. Fruit usually an explosive capsule. *Mainly tropics.*

A large family with over 350 genera and about 4,350 species. Two genera (1 introduced) are found in Europe and 24 in north America. Species of about 40 genera are cultivated, mostly as glasshouse ornamentals, but species of *Acanthus* are generally hardy in Europe.

203 **Pedaliaceae**. Herbs, often sticky-hairy. Leaves opposite/alternate above, simple, exstipulate. Inflorescences racemes/axillary cymes/flowers solitary. Flowers bisexual, zygomorphic. K hypogynous CA perigynous. K(5), C(5), A4/rarely 2, G(2), 2–4-celled; ovules 1–n per cell, axile; style 1, stigma bilobed. Capsule, often 2-horned/nut-like. *Tropics, South Africa.*

There are 18 genera and 55 species. Two genera are native to north America; *Sesamum indicum* is grown as an ornamental and also produces sesame seeds, used in cooking and oil-production.

204 **Martyniaceae**. Herbs/rarely shrubby. Leaves opposite/sometimes alternate above, exstipulate. Flowers in cymes/solitary, bisexual, zygomorphic. K hypogynous CA perigynous. K(5), C(5) 2-lipped, A4 + 1 staminode, G(2), 2–4-celled; ovules n per cell, axile; style 1, stigma bilobed. Capsule, often horned/drupe. *Subtropical America.*

A small family of 5 genera and about 18 species, sometimes included in the *Pedaliaceae*. Four of the genera are native to North America, and species of *Martynia* are occasionally grown for their flowers and hooked fruits.

205 Gesneriaceae. Herbs/shrubs, some epiphytic. Leaves usually opposite/basal, often velvety; ptyxis variable but mainly involute. Inflorescences cymose/flowers solitary. Flowers bisexual, usually zygomorphic. K hypogynous CA perigynous. K(5), C(5), A4/2/rarely 5, G(2); ovules n, parietal, placentas intrusive, bifid (placentation sometimes thus appearing axile); style 1, stigma capitate or slightly bilobed. Capsule/berry. *Mostly tropics*.

A large, mainly tropical family of 146 genera and about 2,400 species. Three genera are native to Europe, 5 to North America. About 46 genera are cultivated as warm-house ornamentals, species of *Saintpaulia* and *Sinningia* (Gloxinia) being the best known.

206 Orobanchaceae. Parasitic herbs, chlorophyll 0. Leaves alternate, scale-like. Flowers bisexual, zygomorphic. K hypogynous CA perigynous. K(4–5), C(5), A4, G(2)/rarely (3); ovules n, parietal, usually on 4 placentas; style 1, stigma 2–3-lobed. *Mainly N temperate areas*.

Lathraea, often placed in the *Scrophulariaceae*, is included here.

There are about 15 genera and 150 species, all parasitic. Four genera occur in Europe and 4 are native to North America.

207 Lentibulariaceae. Herbs, mostly insectivorous, some aquatic. Leaves alternate/basal, often of 2 forms, elaborated, sometimes converted into traps. Inflorescences on scapes, racemes/flowers solitary. Flowers bisexual, zygomorphic. K hypogynous CA more or less perigynous. K2–5/(2–5), C(5) spurred at base, A2, G(2); ovules n, free-central; stigma sessile, 2-lobed or with 1 lobe sometimes reduced. Capsule. *Widespread*.

There are 4 genera and 245 species of plants which trap insects either by sticky hairs or by active traps. Two genera occur in Europe, 2 in North America. Species of *Pinguicula* and *Utricularia* are grown as ornamentals.

208 Myoporaceae. Woody. Leaves usually alternate, often with resinous glands, exstipulate. Inflorescences various. Flowers bisexual, usually zygomorphic. K hypogynous CA perigynous. K(5), C(5),

A4/rarely 5, G(2): ovules 4–8, axile; style 1, stigma capitate. Fruit drupe-like. *Scattered, mostly Australia.*

A small family of 5 genera and about 220 species. A few species of *Myoporum* are cultivated.

209 Phrymaceae. Herbs. Leaves opposite, exstipulate. Inflorescence a spike. Flowers bisexual, zygomorphic, deflexed in fruit. K hypogynous CA perigynous. K(5) teeth hooked, C(5), A4, G(2); ovule 1, basal; style 1, stigma slightly 2-lobed. Nut in persistent K. *E Asia, Atlantic N America.*

A single genus (*Phryma*) with 1 or 2 species, *P. leptostachya* occurring in both North America and the far east.

PLANTAGINALES

Flowers actinomorphic, parts in 4s; C hyaline, small; A exserted; G superior.

210 Plantaginaceae. Herbs/rarely shrublets. Leaves alternate/opposite/basal. Inflorescence usually a spike. Flowers unisexual/bisexual, actinomorphic. K hypogynous CA perigynous/KCA hypogynous. K4/(4), C(3–4), A usually 4, G1–4-celled; ovules few, axile; style 1, undivided. Capsule opening by lid/nut. *Widespread.*

Three genera and about 250 species. Two genera are native to both Europe and north America, and a few species of *Plantago* are grown as ornamentals.

DIPSACALES

Woody/herbaceous; leaves exstipulate; G inferior.

211 Caprifoliaceae. Mostly shrubs/climbers. Leaves opposite, usually simple/rarely pinnate, usually stipulate; ptyxis variable (important in the classification of *Lonicera*). Inflorescence often cymose. Flowers bisexual, actinomorphic/zygomorphic, often twinned. KCA epigynous. K5/(5), C usually (5), A4–5 borne on C-tube, G(3–5) sometimes only 1 cell fertile; ovules 1–n per cell, axile/pendulous;

styles short or long, stigma 3–5-lobed/stigmas sessile, free. Berry. *Widespread, mainly N temperate areas.*

Once thought to be related to the *Rubiaceae* (above), and consisting of 16 genera and up to 400 species. Six genera are native to Europe and 8 to north America. Species of about 10 genera are cultivated as ornamentals, the most important being *Lonicera* and *Viburnum*.

212 **Adoxaceae**. Rhizomatous herbs. Leaves opposite/basal, compound, exstipulate. Flowers in a head, bisexual, actinomorphic. KCA more or less epigynous. K(2–3), C(4–6), A4–6 borne on C-tube, each split into 2 half-anthered parts, G(3–5); ovules 3–5, axile; styles 3–5, short, free. Drupe. *N temperate areas.*

A single genus (*Adoxa*) generally considered to contain only a single species (*A. moschatellina*), though this is divided into 2 or 3 species by some authors. It is native to both Europe and North America.

213 **Valerianaceae**. Herbs. Leaves opposite, simple/dissected, exstipulate; ptyxis conduplicate. Inflorescence cymose. Flowers bisexual, zygomorphic. KCA epigynous. K late-developing, sometimes pappus-like, C(5) often saccate/spurred, A1–4 borne on C-tube, G(3), only 1 cell fertile; ovule 1, pendulous; style 1, stigma 2–3-lobed. Cypsela, K often elaborated in fruit. *Mainly N temperate areas, Andes.*

Seventeen genera and about 400 species. Five genera occur in Europe and 4 in North America. Species of about 6 genera are cultivated as ornamentals.

214 **Dipsacaceae**. Mostly herbs. Leaves opposite, simple/dissected, exstipulate; ptyxis usually conduplicate, rarely involute. Inflorescence an involucrate head, rarely flowers in spiny-bracted whorls. Flowers bisexual, zygomorphic. KCA epigynous with cupular involucel. K5–10/cupular, C(4–5), A4/rarely 2, borne on C-tube, G(2); ovule 1, apical; style 1, often with 2 small stigmatic lobes at apex. *Old World, centred in Mediterranean area.*

There are up to 10 genera and about 250 species. Ten genera

occur in Europe, 6 in North America. Species of about 7 genera are cultivated as ornamentals.

CAMPANULALES

Leaves alternate; A5 rarely −2, often free from C-tube and convergent; G usually inferior.

215 Campanulaceae. Herbs, often with milky sap/rarely woody. Leaves usually alternate, simple, exstipulate; ptyxis variable but mainly supervolute. Inflorescence various. Flowers bisexual, actinomorphic/zygomorphic. KCA epigynous/rarely hypogynous. K5/rarely 3–10, C(5)/rarely (3–10) valvate, A5/rarely 3–10 rarely borne on C-tube, G(2–5)/rarely −(10); ovules n, axile; style 1, stigma shallowly or deeply 3-lobed. Capsule/fleshy. *Widespread.*

A family of 87 genera and nearly 2,000 species. Fifteen genera are native to Europe, 23 to North America, and species of about 20 genera are grown as ornamentals.

Lobeliaceae, with zygomorphic flowers, are sometimes separated as a distinct family.

216 Goodeniaceae. Herbs/shrubs. Leaves usually alternate, simple, exstipulate. Flowers bisexual, zygomorphic. KCA usually epigynous. K5/(5), C(5) 1–2-lipped, valvate/infolded, A5 sometimes borne on C-tube, G(2), 1–2-celled; ovules 1–2 per cell, axile/basal; style 1, stigma sheathed. Fruits various. *Mainly Australasia.*

Sixteen genera and 430 species; 1 genus is native to North America. Species of *Scaevola* are being increasingly grown in Europe as ornamentals.

217 Brunoniaceae. Herbs. Leaves basal, simple, exstipulate. Inflorescence a head with bracts. Flowers bisexual, more or less actinomorphic. K hypogynous CA perigynous. K(5), C(5) valvate, A5 anthers united into tube, G1-celled; ovule 1, basal; style 1, stigma sheathed. Nut enclosed in K-tube. *Australia.*

A single genus (*Brunonia*) with a single species (*B. australis*), occasionally grown as an ornamental in Europe.

218　**Stylidiaceae**. Herbs. Leaves basal/on the stem, linear, usually exstipulate. Inflorescences various. Flowers unisexual/bisexual, actinomorphic/zygomorphic. KCA epigynous. K(5–7), C(5) imbricate, A2 joined to style, G(2); ovules n, axile/parietal/free-central; style 1, with stamens attached, to 1 side of the flower, moving rapidly centtrally when touched at the base. Fruit usually a capsule. *Australasia*.

A small family of 5 genera and 170 species. A few species of *Stylidium* are grown as ornamentals and on account of their touch-sensitive styles.

219　**Compositae/Asteraceae**. Herbs/woody, sometimes with milky sap. Leaves variable, exstipulate, occasionally with small leaflets at the base resembling stipules; ptyxis variable. Inflorescence an involucrate head which is rarely 1-flowered. Flowers unisexual/bisexual, actinomorphic/zygomorphic. KCA epigynous. K reduced to pappus or rarely completely absent, C(5/3), A(5)/rarely (3) anthers joined in a tube, G(2); ovule 1, basal; style 1 at the base, stigma 2-lobed (stigmatic lobes may be longer than style). Cypsela, usually with pappus. *Widespread*.

The largest family of Dicotyledons, with about 1,300 genera and 21,000 species. One hundred and eighty-one genera are native to Europe, 346 to North America. Species of many genera are grown as ornamentals, vegetables and flavourings.

Subclass Monocotyledones

Cotyledon 1, terminal; leaves usually with parallel veins, sometimes these connected by cross-veinlets; leaves without stipules, opposite only in some aquatic plants; flowers with parts in 3s; mature root-system wholly adventitious.

ALISMATALES

Flowers actinomorphic; ovary superior, carpels free; plants aquatic.

[168]

220   **Alismataceae.** Aquatic herbs, often scapose, without latex. Leaves alternate, often broad; ptyxis supervolute. Inflorescence usually much-branched, rarely a single umbel. Flowers unisexual or bisexual, actinomorphic. Perianth 2-whorled; K3, C3, A6–n, G6–n, superior; ovules 1–2 per carpel, basal/marginal. Fruit a group of achenes. *Temperate & tropical areas.*

Thirteen genera and about 90 species of water plants. Six genera are native to Europe, 4 to north America. Species of 5 genera are cultivated as aquatic ornamentals.

221   **Butomaceae.** Aquatics with or without latex. Leaves basal or alternate and borne on the stem; ptyxis supervolute. Flowers solitary or in umbels, subtended by bracts. Flowers bisexual, actinomorphic. Perianth 1- or 2-whorled; P6/K3 C3 persistent or not, A1–n, G6–n, superior; ovules many in each carpel, placentation diffuse-parietal. Fruit a follicle or group of follicles. *Temperate Eurasia, tropics.*

A small family of 4 genera and 13 species. One genus is native to Europe, 2 to North America. Species of 3 genera are grown as ornamental aquatics. The family is often split into 2 segregate families.

1a Leaves linear, latex absent; all perianth-segments petal-like

221a   Butomaceae

b Leaves with stalk and expanded blade, latex present; perianth of 3 sepals and 3 petals   **221b   Limnocharitaceae**

221a   **Butomaceae** in the strict sense. Leaves linear, not clearly divided into stalk and expanded blade; perianth 1-whorled, all segments petal-like. *Temperate Eurasia.*

A single genus (*Butomus*), with a single species, native to Europe and grown there as an ornamental.

221b   **Limnocharitaceae.** Leaves clearly divided into stalk and expanded blade; perianth clearly 2-whorled, outer segments sepal-like, inner petal-like. *Tropics.*

Three genera; species of two of them (*Hydrocleis*, *Limnocharis*) are cultivated as ornamental aquatics.

[169]

222 Hydrocharitaceae. Aquatics usually with at least the flowers emerging from the water. Leaves alternate, variable, usually with stalk and blade. Flowers usually borne in a bifid spathe or between 2 opposite bracts, rarely solitary, unisexual or bisexual, actinomorphic. Perianth in 2 whorls; K3, C3, A1–n, G usually (3–6), inferior; ovules many in each cell, diffuse-parietal. Fruit usually a capsule, rarely berry-like. *Mainly in tropical & warm temperate areas.*

There are 15 genera and about 100 species of submerged, floating or emergent aquatics. Ten genera occur wild in Europe and 10 in North America. Species of 9 genera are grown as ornamentals or as aquarium-oxygenating plants.

223 Scheuchzeriaceae. Herbaceous bog plants. Leaves in 2 ranks, with sheaths bearing ligules. Flowers in racemes, with bracts, bisexual, actinomorphic. Perianth 1-whorled; P6, A6, G3–6, superior; ovules 2 per carpel, basal. Fruit a group of follicles. *Cold N temperate areas.*

A single genus with a single species (*Scheuchzeria palustris*), native to both Europe and North America.

224 Aponogetonaceae. Aquatics. Leaves alternate, long-stalked, with sheathing bases and expanded blades. Inflorescence a simple or forked spike. Flowers usually bisexual and actinomorphic. P1–3, rarely absent, or (when P1) bract-like, A 6 or more, G3–6, superior; ovules few in each carpel, basal. Fruit a group of follicles. *Mainly Old World tropics.*

One genus with about 44 species, 1 or 2 of which are cultivated as ornamental aquatics in Europe.

225 Juncaginaceae. Usually marsh plants, sometimes halophytic. Leaves all basal, sheathing. Inflorescence a raceme or spike, without bracts. Flowers unisexual or bisexual, actinomorphic. Perianth 1-whorled; P6 or rarely 1 (when sometimes interpreted as a bract), A1/4–6, G(3–6), superior, 1-celled; ovules 1 per cell. Fruit a capsule, or indehiscent and of 2 forms. *Mainly temperate & cold regions; Pacific N America.*

Four genera and 18 species, often found in brackish water. One genus is native to Europe.

226 **Potamogetonaceae**. Submerged or emergent aquatics of fresh, brackish or sea water. Leaves alternate or opposite, sometimes in 2 ranks, sometimes with ligules or stipule-like sheath-margins. Inflorescence a spike without bracts or flowers on a flattened axis at first enclosed in a leaf-sheath. Flowers unisexual or bisexual, actinomorphic. Perianth absent or 1-whorled; P0/represented by small lobes or scales/4 (when sometimes considered as swollen connectives of the anthers), A1/3–4, sometimes inserted on the perianthclaws, G4 or 1-celled, superior, stigmas sometimes dilated; ovule 1/1 per carpel, basal or apical. Fruit a drupe/achene/indehiscent. *Widespread*.

In this broad sense there are 7 genera and about 130 species. The family is frequently divided into several segregate families.

1a Perianth of 4 clawed, valvate segments; freshwater aquatics with bisexual flowers in submerged or emergent spikes; carpels 4, free
226a Potamogetonaceae

 b Combination of characters not as above   2

2a Marine plants with densely fibrous rhizomes; leaves mostly basal, with ligules; flowers in stalked spikes subtended by reduced leaves
226d Posidoniaceae

 b Plants marine or in brackish marshes; leaves in 2 ranks or opposite, without ligules; flowers in 2-flowered spikes or on a flattened axis at first enclosed in a leaf-sheath   3

3a Marine plants; flowers unisexual; stigmas not dilated
226c Zosteraceae

 b Plants of brackish marshes; flowers bisexual; stigmas dilated
226b Ruppiaceae

226a **Potamogetonaceae** in the strict sense. Plants of fresh or somewhat brackish water. Flowers bisexual in submerged or emergent spikes. Perianth of 4 clawed, valvate segments. Carpels 4, free. *Widespread*.

Two genera with over 100 species. Both genera are native to

Europe and North America, and a few are grown, mainly as aquarium plants.

226b Ruppiaceae. Plants of brackish marshes. Leaves opposite, without ligules. Flowers sessile in 2-flowered spikes at first enclosed between the bases of opposite leaves, bisexual. Perianth absent. Stamens 2, carpels 4, free, each with a stigma dilated towards the apex. *Temperate & subtropical areas* or on the perianth.

A single genus (*Ruppia*), with about 7 species (sometimes considered to be only a single species). It is native to Europe.

226c Zosteraceae. Marine plants. Leaves in 2 ranks or opposite, without ligules. Flowers unisexual, on flattened axes enclosed by leaf-sheaths. Stigmas not dilated above. *Widespread*.

Three genera and 17 species. One genus is native to Europe, 2 to North America.

226d Posidoniaceae. Marine plants with densely fibrous rhizomes (often washed up on beaches as fibre-balls). Leaves with ligules, mostly basal. Flowers in stalked spikes subtended by reduced leaves, bisexual. Perianth of 4 segments, the 4 stamens borne near their bases (these sometimes regarded as stamens with swollen filaments). *Mediterranean area, Australia*.

There is a single genus (*Posidonia*), with about 3 species; the genus occurs wild in Europe.

227 Zannichelliaceae. Submerged aquatics of fresh or salty water. Leaves alternate/opposite/whorled, entire. Flowers unisexual in axillary cymes or solitary. Perianth cupular/of 3 scales/absent, A1–3, G1–9, superior/naked, stigmas dilated above or 2–4-lobed; ovule 1 per carpel, pendulous. Fruit stalked, indehiscent. *Widespread*.

Four genera with 7 species, 3 genera native to both Europe and North America.

228 Najadaceae. Submerged aquatics of fresh or salty water. Leaves opposite/whorled, entire/toothed. Flowers solitary at the base of the branches, unisexual; male: P2-lipped, A1; female: P membranous/absent, G1-celled, superior/naked, with 2–4 stigmas; ovule 1, basal. Fruit indehiscent. *Widespread*.

A single genus (*Najas*), with 11 species. It is native to both Europe and North America.

LILIALES

Perianth usually petal-like, usually actinomorphic; stamens usually 3/6; G superior/inferior, nectaries often present between the septa on the sides of the ovary or on the perianth.

This large group, together with some others, has been extensively re-classified in recent years, with the recognition of many segregate families, spread over several different orders. The more traditional view given here is more useful for identification purposes. Readers requiring further information should consult Dahlgren, R. T. & Clifford, H. T., *The Monocotyledons: a comparative study* (1982) and Dahlgren, R. T., Clifford, H. T. & Yeo, P. F., *The families of Monocotyledons: structure, evolution and taxonomy* (1985).

229 Liliaceae. Habit diverse: herbs with rhizomes, corms, bulbs, etc., or shrubs or climbers (herbaceous or woody). Leaves all basal or borne on the stem, alternate/opposite/whorled, with main veins usually parallel to margins, sometimes succulent, spiny-margined or rarely reduced to scales, when cladodes present; ptyxis variable, mainly supervolute, rarely flat. Inflorescence usually racemose or umbellate or flowers solitary. Flowers usually bisexual, actinomorphic/weakly zygomorphic. P usually 6/(6)/rarely 3+3, mostly petal-like, A usually 6/(6), G usually (3) superior or inferior; ovules 1–n per cell, axile/rarely parietal. Fruit usually a capsule/berry. *Widespread*.

In this very broad sense, there are about 290 genera and 4,500 species, native to both Europe and North America, many of them also cultivated as ornamentals or vegetables.

The following segregate families are those most likely to be encountered (and found in earlier literature).

1a Leaf-stalk bearing 2 tendrils, or leaf surfaces reversed by a twist in the stalk, or cladodes present in the axils of reduced scale-leaves 2

b Plants without any of the above features   4
2a Ovary inferior; flowers large and showy; fruit a capsule
                                        229c **Alstroemeriaceae**
b Ovary superior; flowers small; fruit a berry   3
3a Plants with ovate, spiny or thread-like cladodes; leaves scale-like
or reduced to small spines   229g **Asparagaceae**
b Shrubs without cladodes; leaves broad, net-veined, their stalks
each bearing 2 tendrils   229f **Smilacaceae**
4a Perianth whorls markedly dissimilar, with parts in 3s or more;
leaves (except for basal scale-leaves) opposite or usually in a whorl
at the top of the stem   229d **Trilliaceae**
b Perianth whorls similar, petal-like, parts usually in 3s; leaves not
as above   5
5a Shrubs or woody climbers with scattered stem-leaves; flowers
solitary, usually pendulous and large; placentation usually parietal;
fruit a berry   229e **Philesiaceae**
b Usually herbs with rhizomes, corms or bulbs, rarely herbaceous or
woody climbers; leaves basal (rarely reduced to sheaths) or those
on the stem spirally arranged or in several whorls, or, if plant
woody and/or with succulent leaves in terminal crowns, then
perianths tubular and sometimes inflated or irregular; flowers
usually with parts in 3s, rarely 2s; placentation usually axile; fruit
a capsule or berry   6
6a Flowers in umbels, each umbel subtended by 1 or more bracts
(spathes) at the base   229b **Alliaceae**
b Flowers not in umbels, bracts various but not as above
                                        229a **Liliaceae**

229a **Liliaceae** in the strict sense. Usually herbs with rhizomes,
corms or bulbs, rarely climbing. Flowers usually in racemes/panicles/
solitary. P usually 6/(6), all usually coloured and generally similar,
A6, G(3), superior/rarely inferior or partly so; ovules 1–n per cell,
axile. Fruit a berry or capsule. *Widespread*.

About 140 genera and over 2,000 species. Many are native to
Europe and North America, and species of many are cultivated as
ornamentals.

Even in this narrower sense, the family is further divided by

Dahlgren and his colleagues, leaving the *Liliaceae* in the strictest sense consisting of the genus *Lilium* and a few allied genera; other families recognised include *Asphodelaceae* (rhizomes, leaves generally narrow, perianth-segments free, fruit a capsule), *Hyacinthaceae* (bulbs, fruit a capsule), *Hostaceae* (rhizomes, leaves conspicuously broadened, perianth-segments united below, fruit a capsule), and *Convallariaceae* (rhizomes, fruit a berry).

229b   **Alliaceae**. Herbaceous plants with bulbs, often smelling of garlic. Leaves mostly narrow, sometimes tubular; Flowers in umbels, each umbel subtended by 2 or more conspicuous bracts (spathes). P usually 6/(6), A6, G(3), superior; ovules 1–n per cell, axile. Fruit a capsule. *Widespread*.

There are about 23 genera and about 750 species. The most important genus is *Allium*, which is native to both Europe and North America, and is widely cultivated both as an ornamental or as a vegetable or flavouring (onion, garlic, leeks, etc.)

229c   **Alstroemeriaceae**. Herbaceous plants with rhizomes. Leaves narrow, surfaces reversed by a twist in the stalk-like base. Flowers generally in umbels without spathes. Perianth of 6 petal-like segments, these distinguished into 2 whorls of 3, the whole perianth slightly zygomorphic. A6, G(3), inferior; ovules axile, n per cell. Capsule. *C & S America*.

Two genera (*Alstroemeria*, *Bomarea*), with almost 200 species, both cultivated in Europe as ornamentals

229d.   **Trilliaceae**. Herbaceous plants with rhizomes. Leaves opposite or more usually in a whorl at the top of the stem. Perianth made up of 2 dissimilar whorls, each of 3 or more segments. G(3–4), superior; ovules many, axile. Fruit a capsule. *N temperate areas*.

There are 4 genera and about 60 species. *Paris* is native to Europe, *Trillium* and *Medeola* to North America. Species of *Trillium* are widely grown as ornamentals.

229e   **Philesiaceae**. Shrubs or woody climbers with scattered, alternate, often rather leathery leaves. Flowers solitary, usually large and

[175]

pendulous. Perianth of 6 similar, petaloid segments. G(3), superior; ovules many, parietal. Fruit a berry. *S hemisphere*.

Eight genera and about 12 species. Species of 4 of the genera are cultivated as ornamentals.

229f **Smilacaceae**. Woody climbers. Leaves alternate, leathery, net-veined, stalked, the stalks each bearing 2 tendrils. Flowers small. P6, A6, G(3), superior; ovules few, axile. Fruit a berry. *Tropical & subtropical areas*.

There are 4 genera and about 350 species. *Smilax* is native to both Europe and North America, and several of its species are culti-vated as ornamental climbers.

229g **Asparagaceae**. Herbs, shrubs or climbers (sometimes woody). Leaves reduced to scales or small spines, their function taken over by variously-shaped cladodes. Flowers borne on the cladodes or in their axils, singly or in clusters. P6/(6), A6, G(3), superior; ovules few, axile. Fruit a berry. *Tropical & temperate areas of the Old World*.

There are 4 genera, all essentially leafless. Species of *Asparagus* are native to Europe, and several genera are found in cultivation, mainly in specialist collections.

230 **Agavaceae**. Woody plants or succulent herbs, often very large. Leaves alternate, usually in basal or terminal rosettes, usually leath-ery, fibrous or succulent, often large and persistent for many years; ptyxis flat or supervolute. Flowers solitary/in racemes/panicles, often produced after prolonged vegetative growth. Flowers usually bisexual, actinomorphic/zygomorphic. P6/(6), A6 usually borne near the top of the perianth-tube, G(3), superior/inferior; ovules n, axile. Fruit a berry or dry and dehiscent. *Mostly subtropical America*.

A family of 18 genera and about 600 species of mainly long-lived, irregularly-flowering succulents. Eleven of the genera are native to North America, and species of *Agave* have become naturalised from cultivation in Europe. Species of about 12 genera are cultivated as small to large succulents.

231   **Haemodoraceae**. Herbs, sap often orange. Leaves mostly basal, often equitant. Flowers in cymes/racemes/clusters, bisexual, actino-morphic/weakly zygomorphic. P6/(6), petal-like, persistent, often densely hairy outside, A6/3, G(3), superior/inferior; ovules n, axile. Fruit a capsule. *Mainly S hemisphere & N America*.

Twenty-two genera with about 160 species. Four genera native to north America, and species of 2 or 3 cultivated as ornamentals.

232   **Amaryllidaceae**. Herbs with bulbs/rhizomes/corms. Leaves usu-ally basal, alternate, often in 2 ranks; ptyxis usually flat, rarely super-volute. Flowers solitary or in umbels, the flower or umbel subtended by 1–several bracts enclosing the inflorescence in bud and usually persisting (spathes). Flowers bisexual, actinomorphic/zygomorphic. P6/(6), A6, anthers sometimes opening by pores, G(3), inferior; ovules 2–n per cell, axile. Fruit a capsule/berry. *Widespread*.

A large family of 70 genera and about 850 species, formerly closely associated with the *Liliaceae*. Seven genera are native to Europe. Species of many genera (especially *Hippeastrum*, *Galanthus* and *Narcissus*) are cultivated as ornamentals.

233   **Tecophilaeaceae**. Herbs with corms or tubers. Leaves linear or ovate. Flowers usually solitary, bisexual, actinomorphic. Perianth with 6 segments united below into a short tube. Stamens 6 or fertile stamens 3, staminodes 3, anthers opening by pores. Ovary half-inferior, 3-celled, ovules numerous, axile. *Subtropical & tropical America, South Africa*.

There are 6 genera and about 20 species. *Tecophilaea cyanocrocus*, from Chile but now apparently extinct there, is widely grown in Europe for the sake of its large, blue flowers.

234   **Hypoxidaceae**. Herbs with rhizomes/corms. Leaves usually basal, alternate, often pleated and hairy; ptyxis conduplicate or pli-cate. Flowers solitary or in racemes or heads, bisexual, actino-morphic. Perianth in 2 series of 3, similar except that those of the outer series are usually hairy on the outside, A6, G(3), inferior; ovules n per cell, axile. Fruit a capsule/berry. *Scattered*.

Often included within the *Amaryllidaceae*, there are 5 genera with

[177]

about 150 species. Species of a few genera (*Curculigo, Hypoxis*) are cultivated as ornamentals.

235   Velloziaceae. Shrubs with forked branches with persistent leaf bases, or woody-based herbs. Leaves alternate, leathery; ptyxis conduplicate. Flowers solitary, terminal, on naked stalks borne in terminal tufts of leaves. P6/(6) petal-like, A6/more in 6 bundles, G(3), inferior; ovules n, axile. Fruit a hard capsule often with spines or glandular. *Tropical Arabia, Madagascar, Africa, S America.*

Two genera (sometimes regarded as several), with about 200 species. A few species are cultivated as glasshouse ornamentals.

236   Taccaceae. Herbs with scapes. Leaves all basal, broad, long-stalked. Flowers in umbels, each umbel with an involucre, inner bracts often dangling. P(6) more or less petal-like, A6, G(3), inferior; ovules n, parietal. Fruit a berry/capsule. *Tropics, China.*

A single genus (*Tacca*), with about 10 species, a few of which are cultivated.

237   Dioscoreaceae. Climbers with swollen rootstocks, sometimes with aerial stem-tubers. Leaves borne on the stems, usually alternate, stalked, often cordate/palmate; ptyxis flat or conduplicate. Flowers in axillary racemes, unisexual, actinomorphic, small. P6/(6) often greenish, A6/3/(6)/(3), G(3), inferior; ovules 2 per cell, axile. Fruit a capsule or berry. *Mainly tropical & warm temperate areas.*

Six genera and over 600 species. Three genera are native to Europe, 2 to North America. Many species are grown as vegetables (yams) in the warmer parts of the world, and a few species are grown as ornamental climbers.

238   Pontederiaceae. Aquatic herbs. Leaves alternate, with sheathing bases, often stalked. Inflorescence racemose, borne in the axil of a spathe-like sheath. Flowers bisexual, actinomorphic/rarely zygomorphic. P(6) petal-like, A usually 6, G(3), superior; ovules 3–n, axile/parietal. Fruit a capsule. *Tropics & warm temperate areas.*

Nine genera, 5 of which are native to North America (4 of them introduced in Europe), and about 30 species of aquatic herbs.

Species of about 5 genera are cultivated, and *Eichhornia crassipes*, which is also cultivated, has become a noxious weed in many tropical river systems.

239 Iridaceae. Terrestrial herbs with rhizomes/corms/bulbs, rarely woody. Leaves alternate, often equitant and/or pleated; ptyxis usually conduplicate or plicate, rarely supervolute. Flowers solitary or in racemes/panicles, bisexual, actinomorphic/zygomorphic. P6/(6) petal-like, sometimes in 2 dissimilar whorls, A3, G(3), inferior, styles 3, often divided; ovules few–n, usually axile. Fruit a capsule. *Widespread, particularly well-developed in South Africa.*

There are about 70 genera and 1,500 species. Nine genera (1 introduced) are found in Europe and 18 are native in North America. Species of many genera (particularly *Iris* and *Crocus*) are cultivated.

JUNCALES

Flowers clustered but not overlapping in spikelets; P6, hyaline/brownish, papery.

240 Juncaceae. Herbs, often rhizomatous, rarely woody. Stems terete in section. Leaves usually basal, spirally arranged, sometimes reduced to sheaths; ptyxis supervolute. Flowers in cymes/panicles/corymbs/heads, usually bisexual, actinomorphic. P6, A usually 6, pollen in tetrads, G(3), superior; ovules 3–n, axile/parietal. Fruit a capsule. *Widespread.*

A family of about 9 genera and 350 species. Two genera are native to both Europe and North America. Many grow in damp or seasonally damp places, and a few species of *Juncus* and *Luzula* are grown as ornamentals.

BROMELIALES

Leaves usually in basal rosettes, stiff and channelled, often with scales all over the surface; perianth strongly differentiated in 2 whorls, the outer whorl sepal-like, the inner petal-like; nectaries frequent in the septa of the ovary.

241 Bromeliaceae. Herbs, often epiphytic. Leaves mainly basal, often spiny-margined and/or with elaborate scales; ptyxis flat or supervolute. Flowers usually in terminal racemes/panicles, with conspicuous bracts, bisexual, usually actinomorphic. K3, C3/(3), A6 anthers usually versatile, G(3), usually inferior; ovules n, axile. Fruit a berry or capsule. *Mainly tropical America*.

A large family of 46 genera and 2,100 species of mainly epiphytic herbs. Ten genera are native to North America. About 80 species belonging to 17 genera are cultivated in glasshouses in Europe.

COMMELINALES

Perianth strongly differentiated into a calyx-like outer whorl and a corolla-like inner whorl; staminodes usually present.

242 Commelinaceae. Terrestrial herbs. Leaves mostly borne on the stems, alternate, with closed basal sheaths; ptyxis involute or supervolute (perhaps helpful in the identification of the genera). Flowers in panicles/coiled cymes or rarely solitary, bisexual, actinomorphic/weakly zygomorphic, K3, C3/rarely fewer, A3–6 anthers basifixed, filaments often hairy, staminodes 0–3, G(3), superior; ovules few, axile. Fruit a capsule. *Tropics & warm temperate areas*.

There are about 35 genera and 600 species. Thirteen genera are native to North America, and species of 3 are naturalised in Europe. A few species are grown as ornamentals and house-plants.

243 Mayacaceae. Aquatic herbs. Leaves alternate, borne on the stems, slender, apices 2-toothed. Flowers axillary, bisexual, actinomorphic. K3, C3, A3, anthers opening by pores, G(3), superior; ovules several, parietal. Fruit a capsule. *Tropical America, Africa*.

A family of a single genus (*Mayaca*) with 4 species, 2 of them native to North America.

244 Xyridaceae. Terrestrial or marsh plants with mostly basal, narrow leaves. Flowers borne in a head, with bracts, bisexual, more or less zygomorphic. K usually 3, segments of differing forms, C(3),

A3 staminodes 0–3, G(3), superior; ovules n–few, parietal. Fruit a capsule. *Mainly tropics.*

A family of 5 genera and 250 species; a few species of *Xyris* are native to North America and a few more are occasionally cultivated there.

245 Eriocaulaceae. Usually marsh plants. Leaves basal or alternate on the stems, sheathing, narrow. Flowers in heads subtended by involucres, unisexual, actinomorphic/zygomorphic. P4–6/(4–6), clearly in 2 whorls of 3 but the segments of each whorl similar, hyaline/membranous, A4–6, G(2–3), superior; ovules 1 per cell, axile. *Mainly tropics.*

There are 14 genera with about 1,200 species, mostly from damp places. One genus is native to Europe and 3 to North America.

GRAMINALES

Flowers overlapping, 2-ranked, in spikelets; P reduced to 2–3 small scales (lodicules); fruit a caryopsis.

246 Graminae/Poaceae. Herbs/bamboos. Leaves in 2 ranks, alternate, sheathing and with ligules; ptyxis conduplicate or supervolute (important in the classification of the genera); stems terete in section, internodes usually hollow. Flowers each compressed between a bract (lemma) and bracteole (palea, rarely absent), the unit comprising a floret, these arranged in 2 ranks in spikelets subtended by 2/rarely 1 empty bracts (glumes). P represented by 2–3 lodicules (often very small), A3/rarely 2/rarely 6 or more, G1-celled, superior, styles 2/rarely 3 or 1; ovule 1, usually lateral. Seed fused to pericarp (caryopsis). *Widespread.*

Economically the most important family of flowering plants, with about 660 genera and 9,000 species. One hundred and fifty five genera are native to Europe and 231 to north America. Many are cultivated as ornamentals, and many are cultivated for their edible grains – wheat (*Triticum*), oats (*Avena*), barley (*Hordeum*), millet (*Sorghum*), rice (*Oryza*), etc.

ARECALES

Palms; usually with woody stems; leaves pleated, splitting into segments or leaflets.

247 **Palmae/Arecaceae**. Trees/shrubs/prickly woody scramblers. Leaves large and pleated, becoming palmately/pinnately divided; ptyxis plicate. Inflorescence a fleshy panicle or spike, often the whole subtended by large, hard, spathe-like bracts. Flowers unisexual/bisexual, actinomorphic. P6/(6), fleshy, A usually 6, rarely more, G usually (3), 1–3-celled; ovules 3; carpels rarely free and with a single ovule each. Fruit a berry/drupe, sometimes huge. *Mainly tropics*.

A family of 212 genera and about 2,700 species of characteristic habit and appearance (palms). Two genera are native to Europe, 16 to North America. Species of about 30 genera are grown as ornamental trees in Europe, particularly in the south.

ARALES

Spathe and spadix usually present; leaves often stalked and broad, net-veined.

248 **Araceae**. Herbs/woody climbers, sap often bitter/milky or rarely the plants reduced, floating aquatics. Leaves usually stalked and broad, often lobed; ptyxis supervolute. Flowers minute, stalkless on a spadix which is enclosed in a conspicuous spathe, unisexual/bisexual, actinomorphic. P4–6/(4–6)/rarely 0, A2–8, G1–n-celled, superior/naked; ovules 1–n, axile/apical/basal/parietal. Fruit usually a berry. *Mainly tropics, less common in temperate areas*.

A large family of 115 genera and 2,000 species. Ten genera are native to Europe, 21 to North America. Species of about 40 genera are cultivated for ornament, and several others are food plants in the tropics.

249 **Lemnaceae**. Small aquatic herbs, often floating, the plant body not differentiated into stem and leaves. Inflorescence minute, in a

pocket on the margin of the plant-body and consisting of 1 or more male flowers, each with 1 or 2 stamens, and a single female flower with a solitary ovary; perianth absent in flowers of both sexes. *Widespread*.

There are 6 genera and 30 species. Three genera are native to Europe, 4 to North America. A few species of *Lemna* are grown in aquaria.

PANDANALES

Woody plants with long stiff leaves; fruits fusing to form syncarps.

250 Pandanaceae. Dioecious trees/shrubs, often with stilt-roots. Leaves crowded, leathery, keeled, often with spiny margins; ptyxis conduplicate. Flowers in panicles on or spadices, unisexual. P rudimentary/0, A n, G1-celled, superior/naked; ovules 1–n, basal/parietal. Fruit a syncarp, the woody or fleshy individual fruits fusing. *Old World tropics, Hawaii*.

Three genera and 675 species of striking woody plants. Two genera are native to north America. Only a few species of *Pandanus* (screwpines) are grown as curiosities in Europe, in large glasshouses.

251 Sparganiaceae. Emergent aquatic herbs. Leaves narrow, alternate. Flowers in unisexual, spherical heads. Perianth of a few scales, A3 or more, G1-celled, superior, more or less stalkless; ovule 1, apical. Fruit drupe-like. *N temperate areas, Australia*.

A single genus (*Sparganium*) with about 12 species; it is native to both Europe and North America.

252 Typhaceae. Marsh herbs. Leaves narrow, alternate. Inflorescence of 2 unisexual, dense, superimposed spikes. Perianth-segments thread-like or of scales, A2–5, G1-celled, superior on a hairy stalk; ovule 1, apical. Fruit dry. *Widespread*.

A single genus (*Typha*), native to both Europe and north America, with about 15 species, occasionally grown for ornament along river- or stream-sides.

CYPERALES

Flowers subtended by membranous bracts arranged spirally/
in 2 ranks in spikes/in spikelets; perianth reduced; fruit
nut-like.

253 **Cyperaceae**. Herbs, sometimes large. Stems terete/3-sided in
section, usually solid. Leaves spirally arranged, with closed sheaths;
ptyxis conduplicate. Flowers subtended by membranous bracts
(glumes) spiral/in 2 ranks in spikes/spikelets without 2 empty
glumes at the base, unisexual/bisexual. P scales/bristle-like hairs,
A6/rarely 3, with basifixed anthers, G1-celled, superior/naked, sometimes surrounded by a flask-shaped structure (utricle); styles 2–3;
ovule 1, basal. Fruit nut-like. *Widespread*.

A large family with 100 genera and 4,000 species. Twelve genera
are native to Europe, 26 to North America. Very few are cultivated
for ornament.

ZINGIBERALES

Leaves usually stalked, with broad blades generally with a main
vein and parallel lateral veins; flowers zygomorphic/asymmetric;
A6–1; G inferior, often with nectaries on the septa.

254 **Musaceae**. Herbs/shrubs, often very large, sometimes leaf-
stalks rolled and forming a stem/trunk-like structure. Leaves spirally
arranged in 2 ranks; ptyxis supervolute. Flowers in coiled cymes or
in the axils of spathes, unisexual/bisexual, zygomorphic. Perianth
usually in 2 series. K3/(3), sometimes joined to the petals or tubular
and split down one side, C3/(3)/2-lipped, A5 sometimes with 1 staminode, G(3) inferior; ovules n or rarely 1 per cell, axile. Fruit a
capsule/schizocarp/fleshy. *Tropics*.

Often divided into segregate families.

1a Leaves and bracts spirally arranged; flowers unisexual; fruit a
   banana   **254a Musaceae**
  b Leaves and bracts in 2 ranks; flowers bisexual; fruit not a
   banana   2

2a Cymes arising from the bases of leaf-sheaths; sepals united below into a long, stalk-like tube; median (upper) petal forming a large lip 254b **Lowiaceae**

b Flowers in coiled cymes in the axils of spathes; sepals free or at most joined to corolla; no petal forming a lip 3

3a Perianth-segments free; ovary with numerous ovules in each cell
254c **Strelitziaceae**

b Perianth-segments partially united; ovary with 1 ovule per cell
254d **Heliconiaceae**

254a Musaceae in the strict sense. Large herbs or apparently small trees, with pseudostems made up of rolled leaf-stalks. Leaves and bracts spirally arranged. Flowers unisexual, in spikes, many in the axil of each of the numerous bracts. Ovules numerous. Fruit a banana. *Old World tropics, introduced in the New World.*

There are 2 genera and about 60 species, including the cultivated banana, *Musa sapientium.* A few species are cultivated for ornament.

254b Lowiaceae. Rhizomatous herbs. Leaves and bracts in 2 ranks. Inflorescences produced directly from the rhizomes. Flowers bisexual. Sepals united below; upper petal forming a large lip. Ovules numerous. Fruit a capsule. *Malaysia to Borneo.*

A single genus (*Lowia*) with about 7 species.

254c Strelitziaceae. Herbs/shrubs with leaves and bracts in 2 ranks. Flowers in coiled cymes in the axes of spathes, bisexual. Sepals free; no petal forming a lip. Ovules numerous, seeds arillate. Fruit a capsule. *Tropics, South Africa.*

There are 5 genera and over 100 species. *Strelitzia reginae*, the bird of paradise flower, is widely cultivated (especially in the south), and is often used as a cut flower.

254d Heliconiaceae. Large herbs with rhizomes. Leaves and bracts in 2 ranks. Flowers in cymes with large and spectacular bracts, bisexual. Sepals free, no petal forming a lip. Ovules 1 per cell. Fruit a capsule. *American tropics.*

A single genus (*Heliconia*) with about 100 species of large, spectacular, tropical herbs. A few species are cultivated in glasshouses in Europe for the sake of their hard, waxy, brightly coloured bracts.

255 Zingiberaceae. Herbs with rhizomes. Leaves in 2 ranks, usually with open sheaths and ligules, aromatic; ptyxis supervolute. Flowers in racemes/heads/cymes, bisexual, zygomorphic. K(3), C3/(3), A1, staminodes usually petal-like, 3–5, lateral staminodes present/absent; stamen modified into a petal-like lip, G(3) inferior, usually surmounted by epigynous glands, style often supported in a groove in the anther; ovules n, axile/parietal. Fruit usually a capsule. *Tropics*.

There are about 40 genera and 1,000 species. Six genera are native to the southern parts of north America. Species of about 17 of the genera are grown as ornamentals, and species of several more are grown in the tropics as spice-plants – ginger (*Zingiber*), cardamom (*Ellettaria*), turmeric (*Curcuma*), etc.

256 Cannaceae. Herbs with rhizomes. Leaves spirally arranged, without ligules; ptyxis supervolute. Inflorescences terminal with flowers in pairs. K3, C(3), A1 petal-like with a half-anther, staminodes several, petal-like, G(3) inferior, style petal-like; ovules n, axile. Fruit a warty capsule. *Tropical America*.

A single genus, *Canna* (which is native to southern North America), with about 50 species, several of which are grown as spectacular, half-hardy bedding annuals.

257 Marantaceae. Herbs with rhizomes. Leaves in 2 ranks, the stalk with a swollen band (pulvinus) at the apex; ptyxis supervolute. Inflorescence a panicle or spike with asymmetric flowers in zygomorphic pairs. K3, C(3), A1 with various petal-like staminodes, G(3) inferior; ovules 1 per cell (2 sometimes aborting). Fruit a capsule, often fleshy. *Tropics*.

There are 32 genera and about 500 species. Three genera are native to North America. Species from 5 genera are grown in glasshouses for the sake of their frequently coloured or marked leaves.

The flowers are asymmetric, but in *Maranta* and other genera occur closely associated in pairs so that the pair of flowers is zygomorphic.

ORCHIDALES

Flowers mostly inverted (resupinate) by means of a twist of 180° in the ovary/flower-stalk, zygomorphic with the median petal usually forming a distinct lip (labellum); A1 or 2; ovary inferior, seeds numerous, with undifferentiated embryos.

258   Orchidaceae. Terrestrial/epiphytic/saprophytic herbs. Leaves alternate/rarely opposite, often borne on succulent, swollen stems (pseudobulbs); ptyxis conduplicate, supervolute or rarely flat (important in the classification of the genera). Flowers solitary or in racemes/panicles, usually bisexual, zygomorphic. K3/rarely (2–3)/2, P3 all petal-like, median usually modified into a lip of varying complexity, A usually 1, rarely 2, united to the style to form a column, pollen in pollinia of varying degrees of complexity, G(3), inferior; ovules very many, parietal. Fruit a capsule. *Widespread.*

Probably the largest family of flowering plants (at least in terms of number of species), with about 800 genera and over 22,000 species. Thirty-five genera are native to Europe and 88 to North America. Many are cultivated, some hardy but most in glasshouses.

Further identification and annotated bibliography

The identification of the family to which a plant belongs is only the first necessary step in its complete identification. To make this book more generally useful some notes are provided on the more important literature which can be used for the purpose. Three broad situations can be defined in so far as identification is concerned. These require somewhat different approaches, and are dealt with separately below. Books and papers are referred to by numbers, and are listed numerically (and alphabetically by author) at the end of the chapter.

The family cannot be satisfactorily identified using the present key. This key does not include all currently recognised flowering plant families. All exclusively tropical and southern hemisphere families have been excluded unless they contain plants widely cultivated in the northern hemisphere; when this is the case, only those plants cultivated are covered by the key, and other members of these families may well not key out. If the family cannot be identified here, several other works may be used. Some of these have keys (6, 13, 15, 17, 24, 26), others are descriptive, often with illustrations (4, 10, 16, 25, 28, 31). A computer based interactive key to the families is published on CD-ROM by CSIRO in Australia (36).

It is important to emphasise here that the circumscription of a particular family can vary from book to book. Care must be taken, therefore, to see that the family arrived at in this book (or any other book) corresponds with the family of the same name in another work. This is often extremely difficult; it may, however, be done by checking the indexes, descriptions, synonyms and comments in the various works against one another. This *caveat* applies to all the works mentioned in this section.

The specimen has been identified to family and its wild geographical origin is known. When the geographical origin of a specimen is known, further identification may be attempted using a Flora of the region or country in question, if one exists. Floras are too numerous to list here, but details can be found in 3 and 12. If no relevant Flora exists, the specimen must be treated as though its geographical origin was not known, as below.

The specimen has been identified to family but its wild geographical origin is not known. Under these circumstances, the first step is to find out if a world-wide monographic study of the family exists. The most notable series of such monographs is that edited by Engler and his successors (9), but this is by no means complete. Other monographic studies are published from time to time in various botanical journals. Most botanical libraries maintain lists of such publications and the *Kew Record* (21) includes short abstracts of almost all current publications.

Attempts to identify the genus to which a plant belongs can be made using various works: 4, 10, 17, 18, 24, 26, which all attempt to be comprehensive (even though some of them are incomplete). 5 and 14 list the names of all current genera, with an indication of the families to which they belong.

If the specimen is from a garden plant, then it may be possible to identify it using a garden Flora; several of these exist: 2, 8, 22, 23, 30, 32, 35.

It is helpful to confirm an identification by comparison with a good illustration. Lists of these can be found in 19 and 34. In recent years, many popular illustrated works on both wild and garden plants have been produced which, though selective and botanically simplified, can be helpful; volume 3 of 35 contains a bibliography in which many such works are listed.

There are several other books which, though not in themselves usable for identification, contain much useful information. Such works include botanical glossaries and dictionaries (1, 7, 11, 20, 27, 28).

Finally, the value of comparing the specimen with named herbarium material cannot be overemphasised. This is the most stringent test of the accuracy of an identification, although it should be mentioned that the naming of herbarium material can be wrong or out-of-date. Use of the herbarium is also helpful when the specimen to be named is too incomplete for identification by means of a key.

Annotated bibliography

1 Airy-Shaw, H. K. (ed.), *J. C. Willis, A dictionary of the flowering plants and ferns*, 8th edn, 1973. A valuable source for plant names and, especially, indication of which genera belong to which families.

2 Bailey, L. H., *A manual of cultivated plants*, 2nd edn, 1949. A detailed account, with keys, descriptions and illustrations, of the 5,000 or so species most commonly cultivated in North American gardens.

3 Blake, S. F. A. & Atwood, A. C., *Geographical guide to the Floras of the world*. Part 1, Africa, Australasia, N & S America and islands of the Atlantic, Pacific and Indian Oceans, 1942; part 2, western Europe, 1961. Lists Floras of the areas mentioned up to the dates specified.

4 Bentham, G. & Hooker, J. D., *Genera Plantarum*, 1862–83. In Latin. Now very out-of-date but still useful for its synopses of genera in each family and its very fine descriptions.

5 Brummitt, R. K. (ed.), *Vascular plants: families and genera*, 1992. An up-to-date, extremely useful listing of families and the genera they include, as recognised at the Royal Botanic Gardens.

6 Cronquist, A., *An integrated system of classification of flowering plants*, 1981. With excellent descriptions and illustrations of the various families; the keys provided are synoptic, that is, they do not take account of all exceptions.

7 Davidov, N. N., *Botanicheskii slovar'*, 1960. A multilingual botanical dictionary (Russian, English, German, French, Latin).

8 Encke, F. (ed.), *Parey's Blumengärtnerei*, 2nd edn, 1958. In German. A taxomonic account, with keys to families and genera, descriptions and illustrations of plants widely cultivated in Germany.

9 Engler, A. *et al.* (eds.), *Das Pflanzenreich*, 107 volumes, 1900–53. In Latin and German. A series of family monographs, with keys, descriptions and illustrations; incomplete.

10 Engler, A. & Prantl, K., *Die Natürlichen Pflanzenfamilien*, several volumes, 1887–99; 2nd edn, several volumes, incomplete, 1924 onwards. In German. With keys, descriptions and illustrations.

11 Featherley, H. I., *Taxonomic terminology of the higher plants*, 1959, facsimile edn, 1965. A standard taxonomic glossary.

12 Frodin, D. G., *Guide to standard Floras of the world*, 1984. The most up-to-date listing of Floras.

13 Geesinck, R., Leeuwenburg, A. J. M., Ridsdale, C. E. & Veldkamp, J. F., *Thonner's analytical key to the families of flowering plants*, 1981.

In English. A very full and complete key covering all flowering plant families. It uses a different taxonomic system and a very different terminology from this book; the 'Introduction and Notes' should be read carefully before attempting the use of the key.

14 Greuter, W. *et al.*, *Names in current use for extant plant genera*, 1993. In English. Lists genus names with indication of the family to which they belong.

15 Hansen, B. & Rahn, K., Determination of Angiosperm families by means of a punched-card system, *Dansk Botanik Arkiv* **26**(1), 1969. A key to families using easily sorted punched cards; the introduction should be read carefully before use.

16 Heywood, V. H. (ed.), *Flowering plants of the world*, 1978 and subsequent reprints. In English. With descriptions, distribution maps and beautiful illustrations for the various families, but no keys.

17 Hutchinson, J., *The families of flowering plants*, 2nd edn, 1959. In English. With keys to families (and sometimes to the genera within them), descriptions and some illustrations; follows an idiosyncratic taxonomic system.

18 Hutchinson, J., *The genera of flowering plants*, vol. 1, 1964, vol. 2, 1967. In English. An attempt to produce an up-to-date Genera Plantarum (see Bentham & Hooker, above), but only two volumes were completed, with keys and descriptions for the genera covered.

19 Isaacson, R. T., *Flowering plant index of illustration and information*, 2 volumes, 1969. References to recent illustrations and articles relevant to garden plants.

20 Jackson, B. D., *A glossary of botanic terms*, 4th edn, reprinted 1953. A standard botanical glossary.

21 *Kew Record of Taxonomic Literature*, 1971 and continuing. Contains abstracts of books and articles of taxonomic interest arranged by families, genera, species, geographical area, etc.

22 Kirk, W. J. C., *A British garden Flora*, 1927. In English. With keys, descriptions and some illustrations to families and genera of plants cultivated in British gardens.

23 Krüssmann, G., *Handbuch der Laubgehölze* 1962, translated into English by Epps, M., as *Manual of cultivated broad-leaved trees and shrubs*, 1986. A very full account of woody plants cultivated in Europe, with some keys, descriptions and illustrations.

24 Kubitzki, K. (ed.), *The families and genera of vascular plants*, vol. 2 (1993). In English. A new attempt to produce a modern Genera Plantarum. Volume 2 contains accounts of 78 small families by

various authors. Volume 1, published in 1990, covers Pteridophytes and Gymnosperms.

25 Lawrence, G. H. M., *Taxonomy of vascular plants*, 1951. In English. With descriptions and illustrations of the families, with much other information, especially references to monographs; without keys.

26 Lemée, A., *Dictionnaire descriptif et synonymique des genres de plantes phanérogames*, 8 volumes, 1925–43. In French. With detailed descriptions of families and genera; keys in volumes 8a and 8b.

27 Mabberley, D. J., *The plant-book*, 1987. A dictionary of information about plants, with brief descriptions, notes and indication for each genus of the family to which it belongs.

28 Melchior, H. (ed.), *Syllabus der Pflanzenfamilien*, 12th edn, 1964. In German. With good descriptions and illustrations of the families, but no keys. Contains much other matter, including lists of important genera in each family, division of the families into subunits (subfamilies, tribes, etc.).

29 Nijdam, J., *Woordenlijst voor de Tuinbouw in zeven talen*, 1952. A polyglot horticultural/botanical dictionary (Dutch, English, French, German, Danish, Swedish and Spanish).

30 Rehder, A., *Manual of cultivated trees and shrubs*, 2nd edn, 1949. The classic account, with keys and descriptions, of woody plants cultivated in North America.

31 Rendle, A. B., *Classification of flowering plants*, vol. 1, 1930, vol. 2, 1938. In English. With descriptions and illustrations of the families and numerous informative notes, but no keys.

32 *The New Royal Horticultural Society Dictionary of Gardening*, 4 volumes, 1992. A dictionary treatment of plants in cultivation; some illustrations, very brief descriptions, occasional keys to species within genera.

33 Schneider, C. K., *Illustriertes Handbuch der Laubholzkunde*, 1904–12. In German. A well-illustrated, very detailed account, with keys, of woody plants cultivated in Europe.

34 Stapf. O. (ed.), *Index Londinensis*, 4 volumes and supplement, 1921– 41. A very complete listing of published plant illustrations up to 1941.

35 Walters, S. M./Cullen, J., *The European Garden Flora*, vol. 1, 1986, vol. 2, 1984, vol. 3, 1989, vol. 4, 1995. A very full treatment, with keys, descriptions and some illustrations of all plants widely cultivated in Europe. Volume 3 contains a bibliography which includes references

to many popular, illustrated books on plants. Volumes 5 and 6 will be published in the near future.

36 Watson, L. & Dallurtz, M. J., *The families of flowering plants – interactive identification and information retrieval*, CD-ROM (1994).

Glossary

Only very brief definitions are given here; if more detail is required, reference should be made to the glossaries cited on pp. 191–193, or to a botanical textbook.

achene: a small dry, indehiscent 1-seeded fruit; in the strict sense, such a fruit formed from a free carpel. See p. 42.

acicular: needle-like.

actinomorphic: regular, radially symmetric, having 2 or more planes of symmetry. See p. 30.

adnate: joined to an organ of another type (e.g. stamens adnate to corolla).

adventitious (of roots): arising from the stem rather than, as is normal, from other roots.

aestivation: the manner in which the perianth parts are arranged relative to each other in bud. See p. 32.

alternate: (of leaves) borne one at each node. See p. 10.

androgynophore: a common stalk bearing the corolla, stamens and ovary above the sepals.

annual: a plant that completes its life-cycle from seed to seed in one year.

antepetalous (stamens): borne on the same radii as the petals or corolla-lobes, and usually of the same number as them. See p. 31.

anther: the pollen-bearing part of the anther, generally made up of 2 or more elongate sacs. See p. 27.

anthophore: a common stalk bearing the stamens and ovary above the calyx and corolla, as in some Caryophyllaceae. See p. 36.

apical (placentation): see p. 27.

apocarpous: having free carpels. See p. 19.

aril: an appendage borne on the seed, strictly an outgrowth of the funicle. See p. 43.

axil: the angle between a leaf and the shoot that bears it.

axile (placentation): see p. 22.

axillary: the adjective from axil (see above).

basal (placentation): see p. 24.

berry: a fleshy, indehiscent fruit with the seeds immersed in pulp. See p. 43.

biennial: a plant that completes its whole life-cycle from seed to seed in 2 years.

bifid: divided into 2 shallow segments.

bilabiate: 2-lipped.

bilaterally symmetric (of a flower or perianth): with a single plane of symmetry. See p. 30.

bipinnate (leaf): a pinnately divided leaf with the leaflets themselves pinnately divided. See p. 12.

biseriate (perianth): with the perianth in 2 whorls (generally sepals and petals).

biserrate: regularly toothed, with the teeth themselves more finely toothed.

blade (of a leaf, petal or sepal): the broad, expanded part, borne on the petiole or claw.

bole: the trunk of a tree.

bract: a frequently leaf-like organ (often very reduced) bearing a flower, inflorescence or partial inflorescence in its axil.

bracteole: a bract-like organ (often even more reduced) borne on the flower-stalk.

bulb: an complex underground storage organ. See p. 8.

caducous: falling off early.

calyptrate (of a perianth) shed as a unit, often in the shape of a cap or candle-snuffer.

calyx: the outer whorl(s) of the perianth, consisting of sepals. See p. 29.

calyx-lobes: the free parts (equivalent to sepals) of a calyx which has a tube at the base.

capitate: head-like.

capitulum: a head-like inflorescence (e.g. dandelion).

capsule: a dehiscent, usually dry fruit formed from an ovary of united carpels. See p. 42.

carpel: the organ containing the ovules; when several are united, they may be much modified and difficult to distinguish. See p. 19.

caruncle: an outgrowth near the point of attachment (*hilum*) of a seed.

caryopsis: an achene with the seed united to the fruit wall. See p. 42.

catkin: a unisexual inflorescence of small flowers with no petals, often deciduous as a whole, and with overlapping bracts. See p. 18.

caudex: the intermediate zone between stem and root. See p. 8.

cauliflory: the bearing of flowers directly on the woody shoots.

cells: the chambers in an ovary of united carpels; also known as *loculi*. See p. 20.

circinate: see p. 14.

climber: a plant which grows on other plants for support.

cladode: a lateral, usually flattened stem-structure borne in the axil of a reduced leaf.

cluster: an indeterminate inflorescence containing several flowers. See p. 18.

collateral (ovules): borne side-by-side.

compound (leaf): divided into distinct and separate leaflets.

compound (fruit): made up of the products of more than one ovary.

conduplicate: see p. 14.

connate: united to other organs of the same type (e.g. petals connate).

connective: the part of the stamen which joins the anther-cells.

contorted: see. p. 32.

coriaceous: leathery and persistent.

corm: an underground stem, very reduced in size and usually vertical.

corolla: the inner whorl of the perianth.

corolla-lobes: the free parts (equivalent to the petals) of a corolla which has a tube at the base.

cordate: heart-shaped.

corona: an outgrowth, usually petal-like, of the corolla, stamens or staminodes.

corymb: a flat-topped raceme.

cotyledon: the first seedling leaf or leaves.

crenate: toothed with blunt or rounded teeth.

cupule: a cup formed from free or united bracts, often containing an ovary.

cyme: a determinate or centrifugal inflorescence. See p. 18.

cypsela: a small, indehiscent, dry 1-seeded fruit formed from an inferior ovary (often loosely termed an achene). See p. 42.

deciduous (leaves): falling once a year; also used of stipules, catkins, etc.

declinate (stamens, styles, etc.): arched downwards and then upwards towards the apex.

decussate (leaves): the opposite leaves of one pair at right angles to those above and beneath it.

dehiscence: the mode of opening of an organ, usually an anther or a fruit.

dentate: toothed.

diffuse-parietal (placentation): see p. 24.

dioecious: with male and female flowers on separate plants.

disc: a fleshy, nectar-secreting organ frequently developed between the stamens and ovary (sometimes also extending outside the stamens).

distichous (leaves): borne alternately on opposite sides of the shoot.

divided (leaves): cut into distinct and separate leaflets (see *compound*).

dorsifixed: attached to its stalk or supporting organ by its back, usually near the middle.

drupe: a fleshy or leathery, 1–few-seeded fruit with a hard inner wall. See p. 43.

drupelet: a small drupe.

elaiosome: an oily appendage borne on a seed, generally near the point of attachment (*hilum*).

endocarp: the inner part of the fruit wall, often hard and stony. See p. 41.

endosperm: storage material found in many seeds, formed after fertilisation and incorporating genetic material from the male parent.

entire (leaves): simple and with unlobed or unthoothed margins.

epigynous: see pp. 32–41.

epigynous zone: see p. 35.

epiphyte: a plant which grows physically on another plant, but is otherwise free-living.

equitant (leaves): folded sharply inwards from the midrib, the outermost leaf enclosing the next, etc.

evergreen (leaves): persisting for more than one growing season.

exfoliating (bark): scaling off in large flakes.

exstipulate: without stipules.

extrorse (anthers): opening towards the outside of the flower.

exocarp: the outer part of the fruit wall, often forming a rind. See p. 41.

false fruit: a fruit which includes tissues not developed from the original ovary.

false septum: a secondary wall (septum) in an ovary cell, formed after the primary walls.

false whorl (of leaves): an apparent whorl of leaves produced by extreme shortening of the internodes between the individual leaves (e.g. *Rhododendron*).

fascicles: an indeterminate inflorescence containing more than 1 flower. See p. 18.

filament: the stalk of the stamen, bearing the anther.

-foliolate: divided into the specified number of leaflets (e.g. 5-foliolate).

follicle: a several-seeded fruit or partial fruit formed from a single carpel, dehiscing along the inner suture.

free-central (placentation): see p. 24.

fruit: the structure containing all the seeds produced by a single flower. See p. 41.

funicle: the stalk of an ovule.

gamopetalous: with the corolla-segments united at the base.

gamosepalous: with the calyx-segments united at the base.

gland: a secretory organ.

glume: see p. 181.

gynaecium: the ovary, the female sex organs of a single flower collectively.

gynoecium: alternative and more usual spelling for *gynaecium*.

gynophore: the stalk of a stalked ovary.

half-inferior (ovary): with the lower part of the ovary below the insertion of perianth and stamens, the upper part above it.

halophytic: growing in saline soils.

half-parasites: plants which have green leaves but which are also parasitic.

hapaxanthic: see p. 6.

head: see *capitulum*.

herbaceous: with the texture of leaves.

herbaceous perennials: a plant dying back to the root at the beginning of each unfavourable season. See p. 6.

hypogynous: see pp. 32–41.

imbricate: overlapping. See p. 32.

imparipinnate: pinnate without a terminal leaflet.

indehiscent (fruit): without any definite opening mechanism.

indumentum: a covering of hair or scales.

inferior ovary: see pp. 32–41.

inflorescence: the arrangement of flowers on a branch or shoot. See p. 16.

internode: that part of a stem between one leaf-base and the next.

intrusive-parietal (placentation): see p. 24.

involucre: a whorl of bracts below an inflorescence.

involute: see p. 14.

laciniate: deeply slashed into narrow segments.

lamina: the broad part of a leaf or petal; see *blade*.

leaflet: the segments of a divided (compound) leaf.

legume: a dry, dehiscent fruit formed from a single carpel, dehiscing along both sutures. See p. 42.

lemma: see p. 181.

lepidote: bearing peltate scales; or such scales themselves.

ligule: a tongue-like outgrowth from a petal or at the junction of leaf-sheath and blade.

locules or *loculi*: the cavity(ies) in a carpel, ovary or anther.

lomentum: an indehiscent fruit which fragments transversely between the seeds, forming 1-seeded segments.

longitudinal dehiscence (of anthers): opening along the length of the anther.

marginal (placentation): see p. 22.

medifixed (hairs): attached by the middle.

mericarp: a 1-seeded portion of a fruit formed from an ovary of united carpels which splits apart at maturity. See p. 42.

-merous: indicating the number of parts (e.g. 3-merous or trimerous).

mesocarp: the central part of the fruit wall, sometimes fleshy.

monocarpic: existing in a vegetative state for several years before flowering.

monoecious: with male and female flowers on the same plant.

multilocular (ovary): with 2 or more cells or loculi.

multiseriate (perianth): a perianth formed from 3 or more whorls of organs.

naked (ovary): see p. 36.

nectary: a nectar-secreting structure, usually within a flower, occasionally on other parts of the plant.

nectariferous disc: a nectar-secreting disc within a flower, generally between stamens and ovary.

node: the point on a stem at which a leaf, a pair of leaves or a whorl of leaves, is attached.

nut: a hard, indehiscent, 1-seeded fruit. See p. 43.

nutlet: a small nut.

obconical: [child's] top-shaped.

obdiplostemonous (flower): with the stamens twice as many as the petals, those of the outer whorl on the same radii as the petals.

opposite (leaves): leaves borne 2 at each node (generally on opposite sides of the stem).

ovules: the structures within the ovary which become the seeds after fertilisation.

palea: see p. 181.

palmate (leaves): divided to the base into separate leaflets, all the leaflets arising from the apex of the stalk. See p. 11.

palmatifid (leaves): divided palmately to about halfway from margin to stalk.

palmatisect (leaves): divided palmately more than halfway from margin to stalk.

panicle: a much-branched inflorescence, strictly a raceme of cymes, but also used for a raceme of racemes.

parallel (veins): veins which are distinct and unbranched from the base of the leaf running parallel towards the apex.

parasitic: a plant which does not photosynthesise (is not green), which receives all its nutrition from the host plant to which it is attached.

parietal (placentation): see p. 22.

paripinnate: a pinnate leaf without a terminal leaflet.

pedicel: the flower-stalk.

peduncle: the stalk of an inflorescence.

peltate: disc-shaped, the stalk arising from the middle of the undersurface.

perianth: the outer, sterile whorls of a flower, often differentiated into calyx and corolla.

pericarp: the wall of the fruit, often differentiated into *exocarp, mesocarp* and *endocarp*.

perigynous: see pp. 32-41.

perigynous zone: see pp. 35.

perisperm: storage tissue in some seeds formed from maternal tissue.

petals: the individual segments of the corolla.

petiole: the leaf-stalk.

petiolule: the stalk of a leaflet.

phloem: tissue within the vascular bundles of the plant which is concerned with the transportation of complex chemicals; only visible with the aid of a microscope.

phyllode: a flattened leaf-stalk which takes the place of a leaf.

pinnate (leaves): bearing separate leaflets on each side of a common stalk.

pinnatifid (leaves) divided pinnately to about halfway from margin to midrib.

pinnatisect (leaves) divided pinnately from halfway from margin to midrib or more.

pistillode: a rudimentary, non-functioning ovary.

plicate: see p. 14.

pluricarpellate (ovary): made up of 2 or more carpels.

pod: a legume.

pollinia: coherent masses of pollen dispersed as units.

polypetalous: with distinct, free petals.

polysepalous: with distinct, free sepals.

pome: a fruit which is made up of an inferior ovary surrounded by fleshy or leathery tissue derived from the receptacle of the flower.

poricidal (anthers): opening by pores.

porose: poricidal.

ptyxis: the manner of folding or rolling of the individual leaves inside the vegetative bud.

pyrenes: the stone(s) within a drupe.

raceme: a simple, elongate inflorescence with stalked flowers, the oldest flowers towards the base. See p. 17.

rachis: the main stalk of an inflorescence or the central axis of a pinnate leaf.

radially symmetric (flower): with several planes of symmetry. See p. 31.

receptacle: the apex of the pedicel where the floral parts are attached.

reticulate (veins): forming a network.

revolute: see p. 14.

rhizome: underground stem bearing scale-leaves and adventitious roots.

root-tubers: see p. 7.

runners: see p. 8.

saccate (perianth or corolla): with a conspicuous, hollow swelling.

sagittate: arrowhead-shaped.

samara: a dry, winged, indehiscent fruit or mericarp, usually 1-seeded.

saprophyte: a plant (often without chlorophyll) which obtains its food materials by absorption of complex organic chemicals from the soil.

scale-leaves: rudimentary leaves borne on a rhizome, or occasionally on stems when true leaves are replaced by *cladodes* or *phyllodes*.

scape: a leafless flower-stalk arising directly from a rosette or basal leaves.

schizocarp: a fruit which splits into separate mericarps.

semi-inferior: half-inferior.

sepal: a free segment of the calyx.

serrate: regularly toothed, saw-like.

sessile: without an obvious stalk.

simple (leaves): not divided into separate leaflets (but possibly lobed).

solitary (flower): one borne singly at the apex of a stem.

spadix: a fleshy spike of numerous small flowers.

spathe: a large bract which subtends an inflorescence.

spike: a raceme-like inflorescence in which each flower is stalkless.

spikelet: a small spike, generally with the flowers more or less enclosed between bracts.

spirally arranged (leaves): arranged one per node spirally along the shoot.

spur (of perianth or corolla): a long, usually nectar-secreting or nectar-holding tubular projection.

stamen: the male sex organ, usually consisting of filament, anther and connective.

staminode: a sterile stamen.

stellate (hair): star-shaped.

stigma: the receptive part of the ovary, generally borne at the end of the style, on which the pollen germinates.

stipule: a pair of lateral outgrowths arising at the point where a leaf is attached to the shoot.

stock: the caudex.

stolon: an overground horizontal stem, generally bearing scale-leaves and rooting at its end.

stones (in fruits): pyrenes.

style: the usually elongate portion of the ovary, bearing the style at its apex.

[203]

subshrubs: having persistent aerial shoots near ground-level.

subulate: needle-like.

suffrutescent: having the character of a subshrub.

superior ovary: see pp. 37–41.

superposed (ovules): borne one above the other.

supervolute: see p. 14.

syncarp: a multiple fruit.

syncarpous (ovary): with the carpels united.

tendril: a touch-sensitive, thread-like organ coiling around objects touched, providing support for climbing plants.

tepals: the distinct segments of a perianth which is not differentiated into calyx and corolla. See p. 29.

terete: circular in section.

testa: the coat of a seed.

tetrads: groups of 4 pollen grains shed as units.

trifid: shortly divided into 3.

trifoliolate: made up of 3 leaflets.

tripinnate: divided pinnately 3 times.

triquetrous: 3-sided.

truncate: ending abruptly, as though broken or cut off.

tuber: a food-storage organ, generally a modified stem or root, borne underground or above ground.

umbel: see a raceme in which the individual flower-stalks all arise from the same point at the top of the inflorescence-stalk.

unicarpellate: made up of a single carpel.

unilocular (ovary): with a single cavity.

uniseriate (perianth): made up of one series of organs.

utricle: a bladdery, indehiscent, 1-seeded fruit.

valvate (sepals or petals): edge-to-edge in bud. See p. 32.

valvular dehiscence (of anthers): opening by flaps or valves.

vascular bundles: tissues concerned with the transport of water and other chemicals, forming a network throughout the plant and particularly conspicuous as the veins in the leaves.

versatile (anthers): pivoting freely on the filament.

[204]

verticillate (inflorescence): the flowers in superposed whorls, each whorl consisting of 2 opposite, often much-modified cymes.

winter-annuals: annuals germinating in the autumn and persisting through the winter as rosettes of leaves.

xeromorphic: with the habit of plants characteristic of arid regions – e.g. with reduced or fleshy leaves, densely hairy, etc.

xylem: woody tissue in the vascular bundles concerned with the transport of water through the plant.

zygomorphic: bilaterally symmetric, having only a single plane of symmetry.

Index